PHOSPHOLIPASE A$_2$
Role and Function in Inflammation

ADVANCES IN EXPERIMENTAL MEDICINE AND BIOLOGY

Recent Volumes in this Series

A Continuation Order Plan is available for this series. A continuation order will bring delivery of each new volume immediately upon publication. Volumes are billed only upon actual shipment. For further information please contact the publisher.

PHOSPHOLIPASE A$_2$
Role and Function in Inflammation

Edited by
Patrick Y-K Wong
New York Medical College
Valhalla, New York

and
Edward A. Dennis
University of California at San Diego
La Jolla, California

PLENUM PRESS • NEW YORK AND LONDON

Library of Congress Cataloging in Publication Data

Symposium on Phospholipase A₂: Pathophysiological Role of Soluble and Membrane-Bound Enzymes (1989: New York, N.Y.)
    Phospholipase A₂.

    (Advances in experimental medicine and biology; v. 275)
    "Based on the proceedings of the Symposium on Phospholipase A₂: Pathophysio-logical Role of Soluble and Membrane-Bound Enzymes, held January 15, 1989, in New York, New York, and the Federation of American Societies for Experimental Biology Symposium on the Release and Function of Phospholipase A₂ from Inflammatory Cells, held March 22, 1989, in New Orleans, Louisiana" — T.p. verso.
    Includes bibliographical references and index.
    1. Phospholipase A₂ — Congresses. 2. Inflammation — Mediators — Congresses. I. Wong, Patrick Y-K II. Dennis, Edward A. III. Federation of American Societies for Experimental Biology Symposium on the Release and function of Phospholipase A₂ from Inflammatory Cells (1989: New Orleans, La.) IV. Title. V. Series.
    RB131.S94   1989   617.2′2   90-7780

ISBN-13:978-1-4684-5807-7   e-ISBN-13:978-1-4684-5805-3
DOI: 10.1007/978-1-4684-5805-3

Softcover reprint of the hardcover 1st edition 1990

Based on the proceedings of the Symposium on Phospholipase A₂:
Pathophysiological Role of Soluble and Membrane-Bound Enzymes,
held January 15, 1989, sponsored by the New York Academy of Sciences, in New York,
New York, and the Federation of American Societies for Experimental Biology
Symposium on the Release and Function of Phospholipase A₂ from
Inflammatory Cells, held March 22, 1989,
in New Orleans, Louisiana

© 1990 Plenum Press, New York
A Division of Plenum Publishing Corporation
233 Spring Street, New York, N.Y. 10013

PREFACE

This volume in the Advances in Experimental Medicine and Biology Series is dedicated to developing an overall view of the "state-of-the-art" of knowledge in the field of phospholipase A_2 and to exploring the role and function of this enzyme in various inflammatory diseases. This book grew out of two major symposia on phospholipase A_2 held in 1989: I). "Phospholipase A_2: Pathophysiological Role of Soluble and Membrane-Bound Enzymes" organized by Dr. Doug Morgan and Dr. Ann Welton of Hoffmann–La Roche and sponsored by the New York Academy of Sciences (January 24, 1989). II). "Release and Function of Phospholipase A_2 from Inflammatory Cells" organized by Dr. Patrick Wong and Dr. Edward Dennis at the FASEB meeting in New Orleans (March 22, 1989).

Readers will find exciting advances in our understanding of the structure, function and molecular biology of phospholipase A_2 research which is presented in this volume. The elucidation of gene structures of phospholipase A_2 should lead to new insights and new approches in the control and regulation of the enzyme activity. This is primary to better understanding the function of phospholipase A_2 and the determination of its activity in cellular and animal systems. Of great importance is the development of animal models for inflammatory disease in order to evaluate the role of phospholipase A_2 and its inhibitors. We are convinced that this volume will be an excellent reference that surveys this field in which numerous investigators are currently actively at work.

The Editors

CONTENTS

MACROPHAGE PHOSPHOLIPASE A_2 ACTIVITY AND EICOSANOID PRODUCTION: STUDIES WITH PHOSPHOLIPASE A_2 INHIBITORS IN P388D$_1$ CELLS

Keith B. Glaser[†], Mark D. Lister[†], Richard J. Ulevitch[*] and Edward A. Dennis[†]

[†]Department of Chemistry, University of California at San Diego, La Jolla, California 92093 and Department of Immunology, Scripps Clinic and Research Institute, La Jolla, California 92037

INTRODUCTION

The inflammatory response involves many different types of tissues and cells; a common modulator produced by many of these cells is the eicosanoids. The eicosanoids (prostaglandins, thromboxanes, leukotrienes, hydroxyeicosatetraenoic acids, lipoxins, etc.) are prominent mediators in the development of inflammatory reactions and have, therefore, been a target for therapeutic regulation. More precisely, the actual target has been the enzymes which control the first step of their biosynthesis from arachidonic acid, the cyclooxygenase and the lipoxygenase.

Another potential point of regulation would be the rate-limiting event in eicosanoid biosynthesis, the release of arachidonic acid from membrane phospholipids. This avenue of regulation is complicated by the lack of detailed information on (a) the activation mechanism for arachidonic acid release, (b) the specific enzymes involved in the release of arachidonic acid and most importantly (c) the absence of specific inhibitors of the enzymes involved in arachidonic acid release, e.g. phospholipase inhibitors. A likely candidate for the release of arachidonic acid would be a membrane associated phospholipase A_2 which can directly hydrolyze arachidonic acid from the sn-2 position of membrane phospholipids. However, it has become apparent that there are other possible pathways in the whole cell which could also provide free arachidonic acid. These mechanisms involve the activation of other phospholipases, most notably a phospholipase A_1-lysophospholipase pathway or a phospholipase C-diacylglyceride lipase pathway, as either pathway could potentially release arachidonic acid from membrane phospholipid stores (1). The determination of the relevant enzyme(s) involved in arachidonic acid release depends upon (a) the isolation and characterization of these enzymes, (b) the evaluation of specific inhibitors of the relevant enzyme *in vitro* (c) the correlation of the effects of these inhibitors on the enzyme *in vitro* with their effects on the production of eicosanoids and arachidonic acid release in the intact cell and (d) the demonstration that the inhibitor of the relevant enzyme(s) has little or no effect on the other enzymes which may be involved in arachidonic acid release (1).

The $P388D_1$ murine macrophage-like cell line (2) has served as a model to begin the elucidation of the relevant enzyme(s) for arachidonic acid release. The phospholipase activities have been characterized in this cell line (3) and several of the enzymes purified and kinetically characterized (4-6). More recently, the membrane associated phospholipase A_2 and prostaglandin production in the intact $P388D_1$ cell have been evaluated with regard to several potential phospholipase A_2 inhibitors, p-bromophenacyl bromide (BPB), dimethyleicosadienoic acid (DEDA), manoalide and manoalogue (7). In this report, we describe some of the kinetic characteristics of the membrane associated, calcium dependent phospholipase A_2, prostaglandin production in $P388D_1$ cells and the evaluation of phospholipase A_2 inhibitors on the partially purified $P388D_1$ phospholipase A_2 and prostaglandin production in the intact cell (7).

SUBSTRATE PREFERENCES OF PHOSPHOLIPASE A_2 IN VITRO

The calcium-dependent phospholipase A_2 isolated from the $P388D_1$ murine macrophage-like cell line has been shown to be a membrane-associated form of phospholipase A_2 (3) and, therefore, would be a likely candidate for the enzyme which could release arachidonic acid from membrane phospholipids. It has been suggested that a substrate-specific form of phospholipase A_2 might exist which would preferentially hydrolyze cellular lipids with arachidonic acid in the *sn*-2 position or *sn*-1 ether-linked phospholipids, alkyl- or alkylenyl phospholipids. Phospholipases have been described which have been suggested to preferentially hydrolyze arachidonyl containing phospholipids (8-10) or alkylenyl phospholipids (11,12). Previous studies with the $P388D_1$ phospholipase A_2 have demonstrated that *in vitro* this enzyme has no preference for the fatty acid present in the *sn*-2 position, palmitate being hydrolyzed equally as well as arachidonate containing phospholipids, or for the polar head group of the phospholipid, e.g., choline or ethanolamine (Table I). The 1-ether-linked phospholipids (alkyl- and alkylenyl- phospholipids) are a major source of arachidonate in human neutrophils (13) and comprise a substantial portion of the phospholipid subclasses in $P388D_1$ cells (14) and other inflammatory cells (11). The high content of arachidonate in the *sn*-2 position of 1-alkyl,2-acyl phosphatidylcholine and the generation of two potent inflammatory mediators upon hydrolysis of arachidonic acid (arachidonic acid for eicosanoid biosynthesis and the formation of lyso-PAF) suggests that these two events may be intimately linked by a specific alkyl-phospholipid hydrolyzing phospholipase A_2 (13). The presence of alkyl phospholipid hydrolyzing phospholipase A_2 has been demonstrated in several cells (8,11) and, therefore, the activity of the $P388D_1$ phospholipase A_2 was examined against alkyl-phospholipids to determine if this enzyme can hydrolyze this substrate and to also determine if this enzyme has a preference for the alkyl- phospholipids over diacyl phospholipids as substrate.

TABLE I

KINETIC CONSTANTS OF MACROPHAGE PHOSPHOLIPASE A_2 TOWARD VARIOUS PC SUBSTRATES[a]

SUBSTRATE	V_{max} (nmol min^{-1} mg^{-1})	K_m (μM)
DIPALMITOYL PC[b]	0.44	1.1
DIPALMITOYL PC[c]	5.1	0.66
1-PALMITOYL, 2-OLEOYL PC[b]	0.20	2.0
1-PALMITOYL, 2-OLEOYL PC[c]	6.3	18
1-STEAROYL, 2-ARACHIDONOYL PC[b]	0.57	2.4
1-STEAROYL, 2-ARACHIDONOYL PC[c]	6.2	13

[a] Adapted, with permission, from Lister *et al.* (5).
[b] Michaelis-Menten portion of the nonlinear regression.
[c] Hill Portion of the nonlinear regression.

The activity of the $P388D_1$ phospholipase A_2 against 1-alkyl-2-oleyl PC and 1-alkyl-2-arachidonoyl PC is shown in Table II. As with the diacyl phospholipids, the $P388D_1$ phospholipase A_2 has two distinct substrate dependence regions (5) between 1 and 1000 μM of alkyl-phospholipid (7). Kinetic analyses best fit a Michaelis-Menten model at the low substrate concentrations ($< 100\ \mu M$) whereas at high substrate concentrations, the activity best fits a Hill model with a Hill coefficient of 2 (7).

TABLE II

KINETIC CONSTANTS OF MACROPHAGE PHOSPHOLIPASE A_2
ACTING ON ALKYL ETHER PHOSPHOLIPIDS[a]

Substrate	Michaelis-Menten		Hill	
	V_{max} (unit[b])	K_m (μM)	V_{max} (unit)	K' (*mM^2)
1-Alkyl,2-oleoyl PC	260	3.9	3,200	46
1-Palmitoyl,2-oleoyl PC[c]	200	2	6,300	18
1-Alkyl,2-arachidonoyl PC	340	2.6	5,100	600
1-Stearoyl,2-arachidonoyl PC[c]	570	2.4	6,200	13

[a] Adapted with permission from Lister *et al* (7).
[b] A unit is a pmol min^{-1} mg^{-1}.
[c] From Lister *et al*. (5).

From these observations, it is apparent that *in vitro* the partially purified
$P388D_1$ phospholipase A_2 has no clear preference for phospholipids with arachidonic acid in the *sn*-2 position, different polar head groups or for *sn*-1 ether-linked
phospholipids (with or without arachidonate in the *sn*-2 position). These types of
generalizations may be regarded as *in vitro* observations and may or may not reflect
the true nature of these phospholipase A_2 enzymes in the cellular environment
where their activity is regulated by the cell and its environment.

EFFECTS OF FATTY ACID ON PHOSPHOLIPASE A_2 ACTIVITY

Although the $P388D_1$ phospholipase A_2 activity is not affected by the fatty
acid in the *sn*-2 position of phospholipids, it is inhibited by unsaturated free fatty
acids (5). The amount of inhibition of phospholipase A_2 activity increases with the
degree of unsaturation of the free fatty acid. Maximum inhibition is observed with
arachidonic acid (C 20:4) with an IC_{50} of 16 μM and a K_I of 5 μM, as determined
from kinetic analyses (5).

It was therefore of interest to evaluate other analogs of natural fatty acids as
inhibitors of the $P388D_1$ phospholipase A_2. As shown in Fig. 1, eicosatetraynoic
acid (ETYA), a dual cyclooxygenase and lipoxygenase inhibitor, or 17-octadecynoic
acid (ODYA), a 5-lipoxygenase inhibitor, had no effect on phospholipase A_2 activity
at concentrations as great as 200 μM. These fatty acid analogs contained triple

bonds rather than double bonds as in arachidonic acid and would therefore assume a more linear conformation similar to that of saturated fatty acids, possibly explaining their lack of inhibitory activity against the $P388D_1$ phospholipase A_2 (7). The fatty acid analog 7,7-dimethyleicosadienoic acid (DEDA) does contain cis-double bonds as in arachidonic acid and does inhibit $P388D_1$ phospholipase A_2 activity in a dose-dependent manner (Fig. 1). The apparent IC_{50} for DEDA was 16 μM, similar to that of arachidonic acid, and almost complete inhibition of phospholipase A_2 activity was observed at 100 μM DEDA. DEDA has been shown to be a weak inhibitor of cellular 5-lipoxygenase (15). It was also found to inhibit cobra venom phospholipase A_2. This fatty acid analog appears to be a potent inhibitor of the $P388D_1$ phospholipase A_2 *in vitro* and its inhibitory activity may be in part due to the cis-conformation of the diene which makes this fatty acid analog more analogous to the cis-tetraene structure found in arachidonic acid.

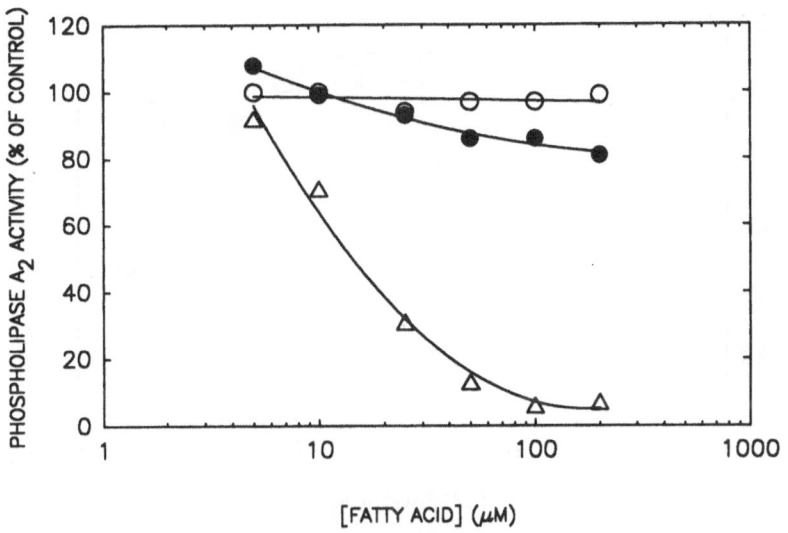

Figure 1. Effect of natural fatty acid analogs on $P388D_1$ phospholipase A_2 activity. Eicosatetraynoic acid (O, ETYA), 17-octadecynoic acid (●, ODYA) and dimethyleicosadienoic acid ($\triangle$, DEDA). Adapted with permission from Lister *et. al.* (7).

EFFECTS OF p-BROMOPHENACYL BROMIDE AND QUINACRINE

p-Bromophenacyl bromide (BPB) is a potent irreversible inhibitor of extracellular venom phospholipases. This reagent alkylates the catalytic histidine residue resulting in complete loss of enzymatic activity for most extracellular venom phospholipase A_2's. BPB was evaluated on the $P388D_1$ phospholipase A_2 and, as shown in Fig. 2, inhibited the phospholipase A_2 activity by only 50% at a concentration of 500 μM BPB, the solubility limit of this compound in aqueous systems. Because the substrate concentration in this *in vitro* assay was 100 μM dipalmitoyl phosphatidylcholine and only 50% inhibition was observed at 500 μM BPB, a 5:1 ratio of inhibitor to substrate, any effect of BPB on the P388D$_1$ phospholipase A_2 enzyme cannot be dissociated from the disruptive effects on the phospholipid substrate, which would also reduce phospholipase A_2 activity. Therefore, this intracellular P388D$_1$ phospholipase A_2 appears to be rather insensitive to inactivation by BPB, which is in contrast to the potent inhibition of extracellular venom phospholipase A_2 by this agent. Another purified macrophage-like cell line (RAW 264.7) intracellular phospholipase A_2 has also been shown to be insensitive to inhibition by BPB, where only 50% inhibition could be obtained at the highest BPB concentrations tested (8).

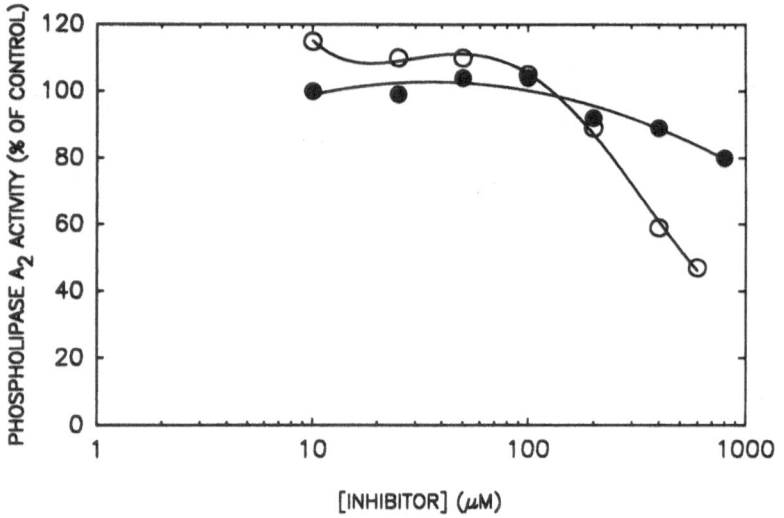

Figure 2. Effect of *p*-bromophenacyl bromide (○) and quinacrine (●) on P388D$_1$ phospholipase A_2 activity. Identical dose response curves were observed with or without preincubation of inhibitor with phospholipase A_2. Adapted with permission from Lister *et. al.* (7).

Another inhibitor used quite frequently in intact cell systems as a phospholipase A_2 inhibitor is the anti-malarial agent quinacrine (mepacrine). We evaluated the effects of quinacrine on the P388D$_1$ phospholipase A_2 and observed no inhibition of phospholipase A_2 activity at concentrations as high as 800 μM (Fig. 2). This is consistent with the lack of inhibitory activity of quinacrine in most *in vitro* assay systems.

MANOALIDE AND MANOALOGUE EFFECTS ON THE P388D$_1$ PHOSPHOLIPASE A$_2$

Manoalide (16) is a marine natural product which has been shown to irreversibly inhibit several different sources of venom phospholipase A$_2$ (17,18). The irreversible inhibition of phospholipase A$_2$ by manoalide is associated with the loss of lysine residues from amino acid analyses (17,19). Mechanistic studies of the inhibition of phospholipase A$_2$ by manoalide have revealed a limited number of lysines modified on bee venom phospholipase A$_2$ suggesting a selective binding site on phospholipase A$_2$ for manoalide (19,20). Manoalogue, a synthetic analog of manoalide, also modifies a limited number of lysine residues on cobra venom phospholipase A$_2$ and retains many of the properties of the parent compound (21). These compounds, manoalide and manoalogue, are potent irreversible inhibitors of extracellular venom phospholipase A$_2$ and were, therefore, examined with the intracellular P388D$_1$ phospholipase A$_2$ to determine if the P388D$_1$ phospholipase A$_2$ was inhibited by manoalide or manoalogue and to evaluate the mechanism of action on the P388D$_1$ phospholipase A$_2$.

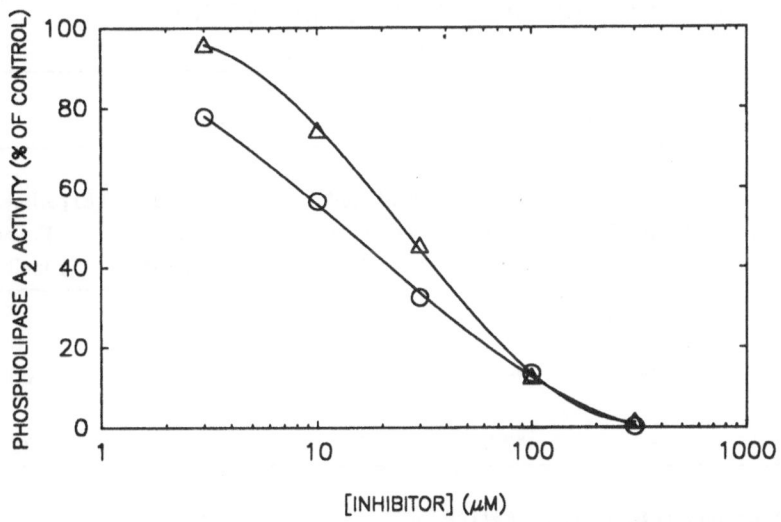

Figure 3. Effect of manoalide (O) and manoalogue (Δ) on P388D$_1$ phospholipase A$_2$ activity. Adapted with permission from Lister *et. al.* (7).

Both manoalide and manoalogue dose-dependently inhibited the P388D$_1$ phospholipase A$_2$ activity as shown in Fig. 3. The apparent IC$_{50}$'s were 16 and 26 μM for manoalide and manoalogue, respectively. Complete inhibition of phospholipase A$_2$ activity was observed at 200-300 μM. Taken together, these results suggest that manoalide and manoalogue are potent inhibitors of the $P388D_1$ phospholipase A_2. It was therefore of interest to determine if their mechanism, i.e. irreversible inhibition, was the same as observed with extracellular venom phospholipase A_2. With the $P388D_1$ phospholipase A_2 it was necessary to compare the inhibition observed after dilution of the preincubation mixture of the $P388D_1$ phospholipase A_2 with manoalide and manoalogue to the inhibition observed by addition of manoalide or manoalogue directly to the assay mixture (preincubation versus dose response). By this comparison, if an inhibitor is irreversible, the dilution of the inhibitor-enzyme complex into the assay mixture following preincubation should result in a level of inhibition (percent) corresponding to the inhibitor concentration in the preincubation mixture. As shown in Table III, upon dilution, the level of inhibition of the phospholipase A_2 by manoalogue corresponds to the assay concentration not the preincubation concentration suggesting a reversible type of inhibition. Some reversal of the manoalide inhibition was also observed but was not as complete as that observed for manoalogue (Table III).

TABLE III

DISTINCTION BETWEEN REVERSIBLE AND IRREVERSIBLE INHIBITION FOR MANOALIDE AND MANOALOGUE[a]

Inhibitor	Concentration			Inhibition	
	Preincubation[b] (μM)	Assay[c] (μM)	Predicted[d] Reversible (% Inhibited)	Predicted Irreversible (% Inhibited)	Experimentally Found (% Inhibited)
Manoalide	30	1.5	<10	70	0
	100	10.0	45	85	60
Manoalogue	250	12.5	25	>95	25

[a] Adapted with permission from Lister et al (7).

[b] Enzyme was preincubated with inhibitor at the designated concentration for 2 hr.

[c] Inhibitor concentration after dilution for assay.

[d] Predicted values were obtained from representative dose response curves.

This observation is in contrast to the observed mechanism of manoalide and manoalogue on extracellular venom phospholipase A_2 which are potently and irreversibly inhibited by these compounds. The reversibility of inhibition of the

P388D$_1$ phospholipase A$_2$ by manoalide and manoalogue is another qualitative difference observed between this particular membrane-associated phospholipase A$_2$ and the more commonly studied extracellular forms of phospholipase A$_2$.

PROSTANOID PRODUCTION AND EFFECTS OF P388D$_1$ PHOSPHOLIPASE A$_2$ INHIBITORS IN THE INTACT CELL

P388D$_1$ cells respond to various stimuli to produce prostaglandins and leuko-trienes (7,22,23). As shown in Table IV, P388D$_1$ cells produce PGD$_2$ as the major cyclooxygenase product with lesser amounts of PGE$_2$, PGI$_2$ (measured as 6-keto-PGF$_{1_\alpha}$) and TXA$_2$ (measured as TXB$_2$) in response to such various stimuli as cal-cium ionophore A23187, melittin and platelet-activating factor (7). The predomi-nance of PGD$_2$ as the major cyclooxygenase product has been observed in several different macrophage-like cell lines, which is unlike peritoneal macrophages where PGE$_2$ and PGI$_2$ dominate the arachidonate metabolite profile (24). This difference has been suggested to be due to the transformed phenotype of the cell lines or to the clonal expansion of a subpopulation of PGD$_2$ producing cells (24). The obser-vation that the P388D$_1$ cells do not produce considerable amounts of TXA$_2$, i.e., amounts of TXA$_2$ are not greater than that of PGE$_2$/I$_2$, suggest that these cells do not resemble the elicited type of macrophage where TXA$_2$ dominates the ara-chidonate metabolite profile (25). The effect of various inhibitors of the P388D$_1$ phospholipase A$_2$ were evaluated in the intact P388D$_1$ cell to determine if they were inhibitory towards the production of PGE$_2$ and the release of [^{3}H]-arachidonic acid from prelabeled cells (in response to calcium ionophore A23187) (7).

DEDA was an effective inhibitor of the P388D$_1$ phospholipase A$_2$ and also in-hibits PGE$_2$ production in P388D$_1$ cells in a dose-dependent manner (IC$_{50}$ $\simeq$ 2.0 μM) as shown in Table V. DEDA has been shown to inhibit release of slow-reacting substance (LTC$_4$, D$_4$ and E$_4$) from rat peritoneal exudate cells but not to affect the ram seminal vessicle cyclooxygenase (15). The relatively high IC$_{50}$ against RBL-1 cell 5-lipoxygenase and an IC$_{50}$ of 16 μM against cobra venom phospholipase A$_2$ suggested that DEDA may be acting in the intact cell by inhibi-tion of arachidonic acid release (15). In P388D$_1$ cells, DEDA was an effective inhi-bitor of PGE$_2$ production; however, DEDA also inhibited the metabolism of exo-genously applied arachidonic acid, suggesting an effect of DEDA on the cellular cy-clooxygenase. DEDA also did not inhibit the release of [^{3}H]-arachidonic acid from prelabeled P388D$_1$ cells (Table V). These results suggest that if DEDA is affect-ing the activity of the P388D$_1$ phospholipase A$_2$ in the intact cell, this effect is dif-ficult to distinguish from the effects this agent has on the cellular metabolism of arachidonic acid.

TABLE IV

PROSTANOID PRODUCTION BY P388D$_1$ MACROPHAGE-LIKE CELLS[a]

Stimulus	Eicosanoid (ng mg protein^{-1})[b]			
	PGE$_2$	PGD$_2$	6-keto-PGF$_{1\alpha}$	TxB$_2$
NONE	3.2	35.7	1.1	N.D.[c]
A23187 (0.5 μM)	15.7	138.2	2.3	0.71
MELITTIN (0.5 μg ml^{-1})	31.5	311.0	1.5	0.62
PAF[d] (1 μM)	10.9	55.7	2.1	0.42

[a] Adapted with permission from Lister *et al.* (7).
[b] P388D$_1$ cells were stimulated for 4 hr at 37 °C and the released arachidonate metabolites were measured by RIA.
[c] N.D. - below detection limit of RIA.
[d] PAF contained 1.0 μM cytocholasin B to prevent chemokinesis;

Manoalide has been shown to inhibit the production of PGE$_2$ in murine resident peritoneal macrophages in response to a variety of stimuli (26) with an apparent IC$_{50}$ in the 0.2 μM range. Manoalide also inhibited the production of LTC$_4$ in response to calcium ionophore cytocholasin B alone had no effect on prostanoid production. A23187 stimulation and [^{3}H]-arachidonic acid release from prelabeled macrophages in response to phorbol esters (26). These data suggested an effect of manoalide on the arachidonic acid release mechanism in resident peritoneal macrophages.

In the macrophage-like P388D$_1$ cell, manoalide and the synthetic analog, manoalogue, dose-dependently inhibited the production of PGE$_2$ in response to calcium ionophore A23187 (Table V). Manoalide or manoalogue had no affect on the metabolism of exogenously applied arachidonic acid at any of the concentrations tested. The effect of manoalide and manoalogue on the release of [^{3}H]-arachidonic acid from prelabeled cells as shown in Table V appears to correlate with the observed inhibition of PGE$_2$ production. These results suggest that manoalide and manoalogue have a direct effect on the arachidonic acid release mechanism in the intact P388D$_1$ cell and this may be correlated with the potent inhibition of the P388D$_1$ phospholipase A$_2$ as observed in vitro.

TABLE V

EFFECT OF CYCLOOXYGENASE, LIPOXYGENASE AND POSSIBLE PHOSPHOLIPASE A_2 INHIBITORS ON PGE_2 PRODUCTION AND [^{3}H]-ARACHIDONIC ACID RELEASE IN P388D$_1$ CELLS[a]

Compound	Concentration (μM)	PGE_2 Production[b] (% of Control)	[^{3}H]-Arachidonic Acid Release[c] (% of Control)
Manoalide	0.5	53	69
Manoalogue	0.5	101	n.d.[e]
	1.0	47	65
	2.0	30	35
DEDA	0.5	120	n.d.
	1.0	75	90
	5.0	38	92
	10	15	88
Indomethacin	0.1	0	n.d.
BW755c	1.0	59	n.d.
	10.0	0	91
PBP[d]	0.5	92	n.d.
	1.0	86	89
	10.0	74	86
Dexamethasone	10	92	n.d.
	100	202	n.d.

[a] Adapted with permission from Lister et al. (7).

[b] PGE_2 production measured by RIA in response to A23187 (0.5 μM) stimulation under conditions in which minimal toxicity was observed. When this was below the level of detection for the RIA, 0% is indicated.

[c] [^{3}H]-Arachidonic acid release measured in the presence of 0.1% BSA (essentially fatty acid free) from 1×10^6 P3888D$_1$ cells prelabeled for 18 hr with [^{3}H]-arachidonic acid and stimulated for 4 hr with A23187 (0.5 μM).

[d] Concentrations of PBP greater than 1 μM were toxic to P388D$_1$ cells.

[e] n.d. - not determined.

The inhibition of the P388D$_1$ phospholipase A$_2$ in vitro by manoalogue was apparently reversible (Table III). In the intact P388D$_1$ cell, the inhibition of PGE$_2$ production was reversed by 44% after a two hour wash period as compared to control (Table VI), suggesting at a least a partially reversible inhibition of PGE$_2$ production in the intact cell. This observation in the intact P388D$_1$ cell also correlates with the observed effects of manoalogue on the P388D$_1$ phospholipase A$_2$ *in vitro* suggesting the possible association of this phospholipase A$_2$ with the arachidonic acid release mechanism in the P388D$_1$ cell.

TABLE VI

REVERSIBILITY OF INHIBITION IN P388D$_1$ CELLS

	% Inhibition of PGE$_2$ Production[a]		
	MLG (1.0 μM)	MLD (0.5 μM)	BW755c (10 μM)
No WASH	78	70	100
2 hr WASH	43	66	11
Δ %	-44	-6	-89

[a] P388D$_1$ cells were preincubated with inhibitor for 2 hr, with or without a 2 hr wash and stimulated for 4 hr with calcium ionophore A23187 (0.5 μM). PGE$_2$ was measured by specific radioimmunoassay. MLG-manoalogue, MLD-manoalide.

CONCLUSIONS

The P388D$_1$ phospholipase A$_2$ is a membrane-associated, calcium dependent enzyme isolated from a macrophage-like cell line (3,4). This phospholipase A$_2$ does not appear to have a preference for the hydrolysis of a specific type of phospholipid, that is, it does not prefer choline over ethanolamine as a phospholipid head group nor does it prefer a specific fatty acid in the *sn*-2 position, e.g. arachidonic acid (5). The P388D$_1$ phospholipase A$_2$ is able to hydrolyze *sn*-1 ether-linked phospholipids, but does not have a strong preference for these phospholipids over the diacyl phospholipids (7). Therefore, *in vitro* the partially purified P388D$_1$ phospholipase A$_2$ appears to have little preference for the phospholipid being hydrolyzed suggesting that, upon activation, this phospholipase A$_2$ would be able to release arachidonic acid, as well as other fatty acids, from a variety of phospholipids available in the intact cell. However, this does not exlude potential regulation of this enzyme in the intact cell which could be interpreted as the preferential hydrolysis of a type of phospholipid or that a different phospholipase A_2 is responsible for stimulated arachidonic acid release in the intact cell.

A variety of potential phospholipase A_2 inhibitors were evaluated on the P388D$_1$ phospholipase A_2. The unsaturated fatty acids are potent competitive inhibitors of the P388D$_1$ phospholipase A_2 (5). Of the analogs of natural fatty acids examined only DEDA was found to inhibit the P388D$_1$ phospholipase A_2 activity (IC$_{50}$ $\simeq$ 16 μM). The presence of the cis-double bonds in DEDA as are found in arachidonic acid may be important in their inhibitory effect on this enzyme as neither saturated fatty acids nor fatty acids with triple bonds (ETYA and ODYA) have a significant effect on the P388D$_1$ phospholipase A_2. The anti-malarial agent, quinacrine, had no effect on the P388D$_1$ phospholipase A_2 consistent with the lack of inhibitory activity of this agent in most *in vitro* assay systems.

An interesting contrast between the P388D$_1$ phospholipase A_2 and extracellular venom phospholipase A_2 is its relative insensitivity to inhibition by BPB. Another phospholipase A_2 from RAW 264.7 cells also appears to be insensitive to inhibition by BPB. Another contrast is the potent inhibition of the P388D$_1$ phospholipase A_2 by manoalide and manoalogue which appears to be reversible in nature as opposed to the potent and irreversible inhibition of extracellular venom phospholipase A_2 by these compounds.

In the intact P388D$_1$ cells, DEDA, manoalide and manoalogue inhibit the production of PGE$_2$. DEDA also appears to affect the metabolism of arachidonic acid in the intact cell making it difficult to separate the potential effects of this agent in the intact cell. However, the effects of manoalide and manoalogue on PGE$_2$ production appear to correlate with the effects observed on the P388D$_1$ phospholipase A_2 *in vitro*, that is, inhibition of PGE$_2$ production appears to correlate with inhibition of [^{3}H]-arachidonic acid release and the effects of manoalogue are at least partially reversible in the intact cell. The production of PGE$_2$ in the P388D$_1$ cells is also insensitive to inhibition by BPB as was the P388D$_1$ phospholipase A_2 *in vitro*. These results suggest the possible association of the P388D$_1$ membrane-associated phospholipase A_2 being studied with arachidonic acid release in the intact P388D$_1$ cell.

These qualitative differences between this phospholipase A_2 and the more commonly studied extracellular venom phospholipase A_2 with respect to the effects of various inhibitors of their activities, suggest caution in the extrapolation of effects of inhibitor on the extracellular type enzymes directly to an intracellular phospholipase A_2, a form of phospholipase A_2 which may be more intimately associated with arachidonic acid release for eicosanoid biosynthesis.

ACKNOWLEDGEMENT

Financial support for the work described in this manuscript was provided by National Institutes of Health Grant GM-20,501 and Lilly Research Laboratories. K.B.G. is a NIH postdoctoral fellow (Grant #HL07926-01).

REFERENCES

1. Dennis, E. A., The Regulation of Eicosanoid Production: Role of Phospholipases and Inhibitors, *Bio/Technology* 5:1294 (1987).
2. Koren, H.S., Handwerger, B.S., and Wunderlich, J.R., Identification of macrophage-like characteristics in a cultured murine tumor line., *J. Immunol.* *114*:894 (1975).
3. Ross, M. I., Deems, R. A., Jesaitis, A. J., Dennis, E. A., and Ulevitch, R. J., Phospholipase Activities of the $P388D_1$ Macrophage-Like Cell Line, *Arch. Biochem. Biophys.* *238*:247 (1985).
4. Ulevitch, R. J., Sano, M, Watanabe, Y., Lister, M. D., Deems, R. A., and Dennis, E. A., Solubilization and Characterization of a Membrane-Bound Phospholipase A_2 from the $P388D_1$ Macrophage-Like Cell Line, *J. Biol. Chem.* *263*:3079 (1988).
5. Lister, M. D., Deems, R. A., Watanabe, Y., Ulevitch, R. J, and Dennis, E. A., Kinetic Analysis of the Ca^{2+} Dependent, Membrane-Bound, Macrophage Phospholipase A_2 and the Effects of Arachidonic Acid, *J. Biol. Chem.* *263*:7506 (1988).
6. Zhang, Y., and Dennis, E. A., Purification and Characterization of Lysophospholipase from $P388D_1$ Macrophage-Like Cell Line, *J. Biol. Chem.* *263*:9965 (1988).
7. Lister, M. D., Glaser, K. B., Ulevitch, R. J., and Dennis, E.A., Inhibition Studies on the Membrane-Associated Phospholipase A_2 *in vitro* and Prostaglandin E_2 Production *in vivo* of the Macrophage-Like $P388D_1$ Cells: Effects of Manoalide, 7,7-Dimethyl-5,8-eicosadienoic Acid, and *p*-Bromophenacyl Bromide, *J. Biol. Chem.* *264*:8520 (1989).
8. Leslie, C.C., Voelker, D.R., Channon, J.Y., Wall, M.M., and Zelarney, P.T., Properties and purification of an arachidonyl-hydrolyzing phospholipase A_2 from a macrophage cell line, RAW 264.7., *Biochim. Biophys. Acta.* *963*:476 (1988).
9. Wijkander, J., and Sundler, R., A phospholipase A_2 hydrolyzing arachidonoyl-phospholipids in mouse peritoneal macrophages., *FEBS* *244*:51 (1989).
10. Okazaki, T., Okita, J.R., MacDonald, P.C., and Johnston, J.M., Initiation of parturition: X. Substrate specificity of phospholipase A_2 in human fetal

membranes., *Amer. J. Obstet. Gynecol. 130*:432 (1978).

11. Loeb, L.A., and Gross, R.W., Identification and purification of sheep platelet phospholipase A_2 isoforms., *J. Biol. Chem. 261*:10467 (1986).

12. Ban, C., Billah, M.M., Truang, T., and Johnston, J.M., Metabolism of platelet-activating factor (1-0-alkyl-2-acetyl-*sn*-glycero-3-phosphocholine) in human fetal membranes and decidua vera., *Arch. Biochem. Biophys. 246*:9 (1986).

13. Chilton, F.H., and Connell, J.R., 1-ether-linked phosphoglycerides: major endogenous sources of arachidonate in the human neutrophil., *J. Biol. Chem. 263*:5260 (1988).

14. Blank, M.L., Smith, Z.L., Lee, Y.J., and Snyder, F., Effects of eicosapentaenoic and docosahexaenoic acid supplements on phospholipid composition and plasmalogen biosynthesis in P388D$_1$ cells., *Arch. Biochem. Biophys. 269*:603 (1989).

15. Cohen, N., Weber, G., Banner, B.L., Welton, A.F., Hope, W.C., Crowley, H., Anderson, W.A., Simko, B.A., O'Donnell, M., Coffey, J.W., Fiedler-Nagy, C., and Batula-Bernardo, C., Analogs of arachidonic acid methylated at C-7 and C-10 as inhibitors of leukotriene biosynthesis., *Prostaglandins 27*:553 (1984).

16. de Silva, E.D., and Scheuer, P.J., Manoalide, an antibiotic sesterterpenoid from the marine sponge *Luffariella variabilis.*, *Tetrahedron Lett. 21*:1611 (1980).

17. Lombardo, D., and Dennis, E. A., Cobra Venom Phospholipase A_2 Inhibition by Manoalide: A Novel Type of Phospholipase Inhibitor, *J. Biol. Chem. 260*:7234 (1985).

18. Glaser, K.B., and Jacobs, R.S., Molecular pharmacology of manoalide: inactivation of bee venom phospholipase A_2., *Biochem. Pharmacol. 35*:449 (1986).

19. Glaser, K.B., Vedvick, T.S., and Jacobs, R.S., Inactivation of phospholipase A_2 by manoalide: localization of the manoalide binding site on bee venom phospholipase A_2., *Biochem. Pharmacol. 37*:3639 (1989).

20. Glaser, K.B., and Jacobs, R.S., Inactivation of bee venom phospholipase A_2 by manoalide: a model based on the reactivity of manoalide with amino acids and peptide sequences., *Biochem. Pharmacol. 36*:2079 (1988).

21. Reynolds, L. J., Morgan, B. P., Hite, G. A., Mihelich, E. D., and Dennis, E. A., Phospholipase A_2 Inhibition and Modification by Manoalogue, *J. Am. Chem. Soc. 110*:5172 (1988).

22. Nitta, T., and Suzuki, T., F$_c$ γ_{2b} receptor-mediated prostaglandin synthesis by a murine macrophage cell line (P388D$_1$)., *J. Immunol. 120*:2527 (1982).

23. Aussel, C., and Fehlman, M., α-fetoprotein stimulates leukotriene synthesis in P388D$_1$ macrophages., *Cell. Immunol. 101*:415 (1986).

24. McGuire, J.C., Richard, K.A., Sun, F.F., and Tracey, D.E., Production of

prostaglandin D_2 by murine macrophage cell lines., *Prostaglandins 30*:949 (1985).

25. Tripp, C.S., Leahy, K.M., and Neeleman, P., Thromboxane synthase is preferentially conserved in activated mouse peritoneal macrophages., *J. Clin. Invest. 76*:898 (1985).

26. Mayer, A.M.S., Glaser, K.B., and Jacobs, R.S., Regulation of eicosanoid biosynthesis *in vitro* and *in vivo* by the marine natural product manoalide: a potent inactivator of venom phospholipases., *J. Pharm. Exp. Ther. 244*:871 (1988).

LOCALIZATION AND EVOLUTION OF TWO HUMAN PHOSPHOLIPASE A$_2$ GENES AND TWO RELATED GENETIC ELEMENTS

Lorin K. Johnson[*], Susan Frank[1], Peter Vades[+], Waldemar Pruzanski[+], Aldons J. Lusis[1], and Jeffrey J. Seilhamer[#]

[*]Salix Pharmaceuticals, Inc. 1507 Kennewick Drive, Sunnyvale, CA 94087, [+]The Wellesley Hospital, University of Toronto, Toronto, Ontario Canada M4Y 1J3, [1]the Depts. of Medicine and Microbiology and Jonsson Comprehensive Cancer Center, UCLA, Los Angeles, CA 90024, and [#]Ideon Corporation, 515 Galveston Drive, Redwood City, CA 94063.

SUMMARY

Mammals are now known to contain at least two distinct classes of phospholipases A$_2$, the progenitors of which can be seen in the venoms of snakes. Mammalian "Type I" PLA$_2$, synthesized primarily by the pancreas, is also present in smaller amounts in other tissues including lung, spleen, and kidney. Recently, a mammalian "Type II" PLA$_2$ has been sequenced, and shown to occur in platelets, synovial cells and fluid, cells of inflammatory peritoneal exudate, liver, intestine, kidney, and placenta. This form, referred to here as Type IIA PLA$_2$, could play a key role in arachidonate release in both normal and pathologic inflammation. The genes encoding both forms have also been recently cloned. Here, the sites of synthesis and respective roles of the two known enzymes are discussed, along with an analysis of the evolutionary conservation of Type IIA PLA$_2$ gene sequence. In addition, two related genetic elements containing sequences homologous to a portion of Type II PLA$_2$ are described, which map to the same chromosome as the Type IIA PLA$_2$ gene (chromosome 1). Either or both of these could also encode a portion of additional mammalian PLA$_2$s.

INTRODUCTION

Phospholipase A_2 (PLA_2) is an important and abundant enzyme present in the digestive secretion of the pancreas of most mammals and in the venoms of snakes and bees. It also occurs in virtually all cell types and participates in such diverse and specialized functions as membrane remodeling[1], digestion of phagocytosed bacteria[2], metabolism of pulmonary surfactant phospholipids[3], and intestinal digestion of dietary phospholipids[4]. Another well-documented role of PLA_2 is the initation of the release of fatty acid precursors of inflammatory eicosanoids[5,6]. It is well documented that high levels of PLA_2 are found extracellularly in inflammatory exudates such as synovial[7,8] and peritoneal fluids[9] from patients with arthritis and peritonitis and in the plasma of patients with acute sepsis[10,11,12]. The relative roles of secreted versus intracellular membrane-bound PLA_2 in the latter processes remain controversial. Nevertheless, a causative role in inflammation for the secreted form of PLA_2 is further implied from studies in which the injection of purified synovial fluid PLA_2 into knee joints of experimental animals resulted in inflammatory and proliferative changes in the synovium[13].

These findings have increased the need for knowledge of the PLA_2 enzymes actually involved in these processes, as well as an understanding of those which are not. Important information includes enzymology and primary sequence data, in addition to an understanding of the regulation of these genes. Ultimately, such efforts may lead to the ability to produce recombinant supplies of inflammatory PLA_2 for inhibitor screening and design. While the enzymology and protein biochemistry of pancreatic PLA_2s have been extensively studied[14-17], relatively little is known of the other mammalian PLA_2 forms. Until recently, even less was known about the structure and regulation of the genes encoding these important enzymes. Here we review the current knowledge of the genes encoding mammalian PLA_2s, map their human chromosomal locations, and examine the degree to which these genes are conserved in other mammalian species.

MATERIALS AND METHODS

The isolation and cloning of the human genes and cDNAs encoding Type I (pancreatic) PLA_2 and Type IIA (synovial fluid peak A; RASF-A) were described elsewhere[23,36]. The cloned cDNAs of both enzymes were used as hybridiziation probes against blots of genomic DNA isolated from the indicated species. Ten ug of genomic DNA from each species was digested with Eco RI overnight, then electrophoresed on 0.8% agarose gels. The ethidium bromide-stained gel was denatured and neutralized by standard methods, blotted overnight onto Hybond nylon membranes (Amersham Inc., Arlington

Heights, IL), and the DNA was subsequently cross-linked using a Stratalinker ultraviolet lamp (Stratagene, San Diego, CA). The filters were prehybridized at 37°C in 5 X SSC (1X SSC is 1.0 mM NaCl, 0.1mM Na citrate), 5 X Denhardt's solution, 50 mM $NaPO_4$, and 10 ug/ml sheared DNA, and were hybridized in the same solution plus 10^7 cpm radiolabelled hybridization probe. The filters were subsequently washed in 1X SSC, 0.1% sodium dodecyl sulfate for one hour at the indicated temperatures. Hybridization probes were prepared from subclones of genomic DNA containing the appropriate gene by polymerase chain reaction (PCR)[48], followed by purification by agarose gel electrophoresis. The isolated DNA was then radiolabelled with ^{32}P by random-primed synthesis and unincorporated nucleotides were removed by chromatography on DE52 (Whatman BioSystems Ltd., Maidstone, Kent, England).

For human gene chromosomal localization, DNA hybridization probes were made from unique restriction fragments containing the gene regions described below ("Homologous PLA_2 Gene Elements"), avoiding repetitive DNA segments. The probes were hybridized against blots of genomic DNA from a panel of somatic cell hybrids derived from the fusion of thymidine kinase-deficient mouse cells and normal human male fibroblasts (IMR91) as described elsewhere[47].

RESULTS AND DISCUSSION

A comparative alignment of the venom PLA_2s of reptiles and insects[14-15,18] has led to the subdivision of the enzymes into two families by virtue of their disulfide bonding patterns. "Type I" PLA_2s, are found in the "old world" snake venoms of elapid variety. The "Type II" PLA_2's found in "new world" snakes of the crotalid and viperid varieties, are easily distinguished by the translocation of paired half-cysteines from positions 11 and 77 to 50 and 132, and also contain seven additional residues at the carboxy-terminus. The two classes of enzymes tend to have different spectra of pharmacological properties[19]. Snake venoms usually contain one or the other type, depending upon their phylogenetic origin (see Refs. 14,15). This structural classification is now relevant to mammalian PLA_2's, since it is clear that mammals contain at least one form of each enzyme type.

Mammalian Type I PLA_2

Pancreatic PLA_2's have now been isolated and sequenced from numerous mammals[14-15,20-23,39]. All known pancreatic PLA_2's contain 14 half-cysteine residues organized in the Type I pattern. Their strong evolutionary conservation is evident in the alignment of three such sequences shown in Fig. 1. Unlike the case in porcine[39] and bovine[39] pancreas, in which isoenzymic forms of pancreatic PLA_2 have been isolated and sequenced, human pancreatic PLA_2 appears to be encoded by a single 4.9 kb gene per haploid human genome[23]. This gene has been isolated, subcloned, and sequenced[23]. The 126-amino acid coding region of fully-processed PLA_2 was contained in three exons of almost equal size, and a membrane translocation peptide was partially encoded in an additional upstream exon.

MAMMALIAN PLA_2 AMINO ACID SEQUENCES

Exon 2:	Type	1 10 20 30 40
porcine	I	ALVQFRSMIKCAIPGSHPLMDFNNYGCYCGLGGSGTPVDELDR
rat	I	AVVQFRNMIKCTIPGSDPFREYNNYGCYCGLGGSGTPVDDLDR
human	I	AVVQFRKMIKCVIPGSDPFLEYNNYGCYCGLGGSGTPVDELDK
		* * *** **
human	IIA	NLVNFHRMIK-LTTGKEAALSYGFYGCHCGVGGRGSPKDATDR
rat	IIA	SLLEFGQMIL-FKTGKRADVSYGFYGCHCGVGGRGSPKDATDE
porcine	IIA	DLLNFRKMIK-LKTGKAPVPNYAFYGCYCGLGGKGSPKDATD?
rabbit	IIA	HLLDFRKMIR-YTTGKEATTSYGAYGCHCGVGGRGAPK?A

Exon 3:		44 50 60 70 80 85
porcine	I	CCETHDNCYRDAKNLDSCKFLVDNPYTESYSYSCSNTEITCN
rat	I	CCQTHDHCYNQAKKLESCKFLIDNPYTNTYSYKCSGNVITCS
human	I	CCQTHDNCYDQAKKLDSCKFLLDNPYTHTYSYSCSGSAITCS
		**
human	IIA	CCVTHDCCYKRLEKR-GC-----GTKFLSYKFSNSGSRITC-
rat	IIA	CCVTHECCYNRLEKS-GC-----GTKFLTYKFSYRGGQISCS
porcine	IIA	CCAAH
rabbit	IIA	KFLSYKFSMK

Exon 4:		86 90 100 110 120 130
porcine	I	SKNNACEAFICNCDRNAAICFSKAPYNKEHK-NLDTKKYC
rat	I	DKNNDCESFICNCDRQAAICFSKVPYNKEYK-DLDTKKHC
human	I	SKNKECEAFICNCDRNAAICFSKAPYNKAHK-NLDTKKYCQS
		**
human	IIA	AKQDSCRSQLCECDKAAATCFARNKTTYNKKYQYYSNKHCRGSTPRC
rat	IIA	TNQDSCRKQLCQCDKAAAECFSRNKKSYSLKYQFYPNKFCK??TPSC
rabbit	IIA	KAAAACF QFYPANRCSGRPPSC

Figure 1. Types I and IIA PLA_2 sequence alignment. The reported amino acid sequences of porcine (Ref. 20), rat (Ref. 21), and human (Refs. 22 and 23) pancreatic Type I PLA_2 were aligned with sequences reported for Type IIA PLA_2 from porcine intestine (Ref. 24), rat platelet (Refs. 25 and 26), spleen (Ref. 28), peritoneal exudate (Ref. 29), and liver mitochondria (Ref. 30), rabbit ascites fluid (Ref. 31) and platelet (Ref. 33), human synovial fluid (Refs. 34-37), platelet (Ref. 37), and placenta (Ref. 38). Residue 114 of the rat platelet sequence (Pro) might also be a Leu (Ref. 27), and residue 1 of the rat liver mitochondrial sequence was reported as Asp (Ref. 30). Residues 27 and 115 of rabbit ascites (Ser and Asn) appeared as His and Ala in the rabbit platelet PLA_2 (Ref. 33). Residue 1 of rabbit neutrophil PLA_2 appears as an Ala (Ref. 32). (*) denotes residues of known catalytic or calcium binding involvement (Refs. 14-17).

Human Type I PLA$_2$ is clearly produced in or translocated to other tissues besides pancreas. For reasons of tissue availability, we obtained a cDNA encoding Type I PLA$_2$ from lung tissue[23], suggesting its synthesis there or in cells contained therein. In addition, Type I PLA$_2$ has been isolated from rat spleen[40] and gastric mucosa[41]. A recent report detected immunoreactivity to anti-pancreatic PLA$_2$ in a transformed fibroblast cell line[43]. Other recent studies have detected immunoreactive pancreatic PLA$_2$ in human lung, stomach, small intestine, and kidney[42]. Therefore, while the functions of the Type I PLA$_2$ gene product in these cases have not been elucidated, it is clear that transcription of this gene is not limited to pancreas in mammals.

Mammalian Type II PLA$_2$

The amino acid sequences published thus far for non-pancreatic mammalian PLA$_2$s include both secreted and membrane-bound forms of the enzyme. In some of these instances the complete amino acid sequence is available from actual sequencing of the protein or from cDNA and genomic cloning. In other cases only partial sequences are available. However, even in these later cases enough information is available to determine the identity of certain PLA$_2$ forms from both secreted and membrane associated sources. Shown in Fig. 1 is an alignment of the reported primary amino acid sequences of non-pancreatic PLA$_2$s isolated from rat, rabbit, porcine, and human sources.

In the human case, secreted forms from synovial fluid[34-38], platelets[37], ascites (peritoneal) cells[36], and placenta[38], have been shown by four different groups to be identical to the sequence shown in Fig. 1. Similarly, in rat, the membrane and secreted forms from platelet[25,26], spleen membrane fraction[28], peritoneal exudate[29] and liver mitochondria[30] are so far identical in sequence (except for the amino-terminal residue of the rat liver mitochondria sequence[30] and conflicting data for residue 114 of the platelet sequence[27]). In rabbit, separate PLA$_2$ isolates from ascites fluid[31], neutrophils[32], and platelet secreted and membrane-bound forms[33] differed only at the amino-terminal residue and two internal residues (see legend, Fig. 1). Therefore, so far the Type II PLA$_2$ sequences from various sources within the same species appear to be very similar, if not identical. Whether the few observed differences within species reflect differences in individuals, or are the result of post-translational modification needs to be resolved in each case.

The similarity of overall amino acid sequence between these enzymes from different species suggests that they represent homologous enzyme forms (72.5% amino acid homology between rat and human).   The identical cysteine structure, active site, and Ca^{++}-binding loop appear in all of these sequences.   Accordingly, we have henceforth referred to this class of probable enzyme homologs as Type IIA PLA_2, under the expectation that other Type II mammalian enzymes may exist (e.g. see below).   It also appears quite likely that the PLA_2 isolated from porcine ileum, by virtue of its sequence homology but in spite of its different tissue of origin, constitutes porcine Type IIA PLA_2.   It is, however, worthy of note that several regions (positions 15-21, 72-82, 85-97, and 110-133 in Fig. 1) are quite different in sequence between the species, although these are also regions of lesser conservation among other PLA_2 sequences[14,15,39].

The gene encoding human Type IIA PLA_2 has recently been cloned and sequenced[36,37].   The gene spans 4.5 kb and is contained in five exons.   The latter two introns divide the 124 amino acid sequence encoding the fully processed PLA_2 protein at similar locations as they do in Type I PLA_2.   The sequence lacks the propeptide seen in the Type I PLA_2 gene, but contains a membrane translocation signal of 20 residues preceeding the enzyme sequence. Transcription of the gene was detected in cells from a peritoneal lavage and synovial tissue[36], and in another study in tissue from tonsil, placenta, and kidney[37].

Several tissues and extracts have been shown to contain both Type I and Type IIA enzyme forms.   A 108,000 X G supernatant solution from homogenized rat spleen contained a PLA_2 identical in sequence to the pancreatic form, constituting 10% of the total cellular PLA_2 activity[40].   From the pellet of the same extract, a PLA_2 enzyme identical in sequence to rat Type IIA PLA_2 was isolated, constituting the remaining 90% of the total activity present[28].   Human kidney has been shown to contain both pancreatic PLA_2 immunoreactivity[42] and transcripts of Type IIA PLA_2 mRNA[37]. Pancreatic, in addition to at least one non-pancreatic, PLA_2 have been detected immunologically in human serum[11,42,44].   Therefore, the systemic distribution of these two enzyme forms may be compartmentalized in some cases and mixed in others.

Interspecies Conservation of PLA_2 Gene Sequences

Type I PLA_2s appear to maintain a very high level of interspecies conservation (Fig 1).   While a sequence is not available for rabbit pancreatic PLA_2, comparison between the human, rat and porcine enzymes shows a 79% identity of amino acid sequence between the three species.   These genes are so conserved at the DNA level that we were able to detect strongly hybridizing bands (i.e. putative venom and/or PLA_2 genes) by hybridization of pancreatic PLA_2 cDNA probes to blots of rat, canine, and porcine as well as the relatively evolutionarily distant cobra (Naja naja), genomic DNA[23].

We extended this hybridization analysis in a similar fashion to the Type IIA gene, to examine whether homologous sequences could be detected within the genomes of other related species. Nylon filter blots of agarose gels containing Eco RI-predigested genomic DNA from several representative species were hybridized with a probe prepared by PCR containing exons 2, 3, and 4 of the Type IIA PLA$_2$ gene. The filters were probed at moderate hybridization (50% formamide, 5X SSC, at 37°C) and low wash (50°C, 1X SSC) stringency. The results are shown in Fig. 2, panel A. A single strongly hybridizing band was detected within the lane containing human DNA, consistent with the presence of a single gene. Single strongly hybridizing bands were also present in the lane containing monkey and porcine genomic DNA, suggesting the DNA sequence of monkey and porcine Type IIA PLA$_2$ may be very similar to the human form. In the lanes corresponding to all of the other mammals, two or more lighter, cross-hybridizing bands were

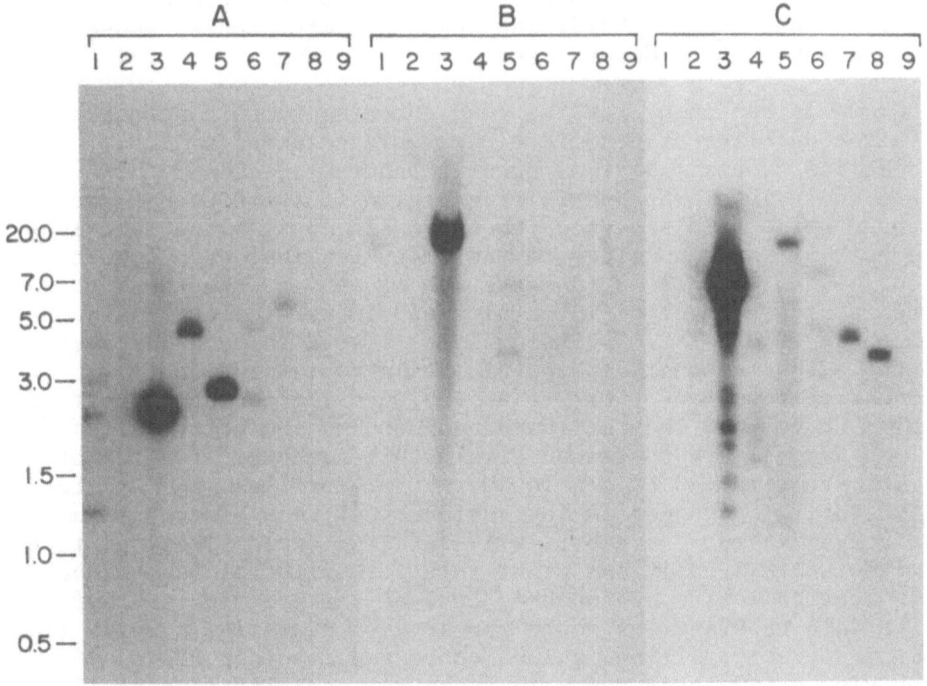

Figure 2. PLA$_2$ gene zoo blots. Genomic DNA extracted from the indicated species was digested with Eco RI, electrophoresed on an 0.8% agarose gel, blotted to nylon filters, and hybridized with ^{32}P-labeled DNA (Ref. 36) as described in Methods and Materials. Panel A, hybridized with a probe comprising exons 2, 3, and 4 of the gene encoding Type IIA PLA$_2$, washed at 50°C; Panel B, an identical blot hybridized with a probe comprising the coding region from clone 8 and 200 bp upstream and downstream flanking sequence, washed at 55°C; Panel C, an identical blot hybridized with a probe comprising the coding region from clone 10 and 200 bp upstream and downstream flanking sequence, washed at 50°C. Lanes 1-9 correspond to DNA samples from sheep (1), dog (2), human (3), pig (4), monkey (5), rabbit (6), mouse (8), and cobra (9). Fragment sizes are given in kilobases.

detected. Possibly, the hybridizing sequences within these genomes contain an internally located Eco RI site. Alternatively, these mammals could contain two or more cross-hybridizing genes. Thus, it is apparent that, as with Type I PLA_2, sufficient homology exists between the mammalian species represented to allow detection by cross-hybridization.

Homologous PLA_2 Gene Elements

In light of the essential and diverse processes in which PLA_2 is known to function, one might have expected to find multiple enzyme forms of PLA_2 encoded by a family of homologous but discrete genes. It now appears, however, that both of the two known PLA_2 genes are present in mammalian genomes in a single copy and do not appear to be a member of a large gene family. Nevertheless, during the course of cloning the two known PLA_2 genes, we were able to detect two other cross-hybridizing human genomic clones, whose function has yet to be determined.

While searching for the gene encoding human pancreatic PLA_2, a second class of genomic clones were detected by hybridization probes made from porcine pancreatic cDNA[23]. DNA sequence analysis of the hybridizing region of these clones yielded an encoded amino acid sequence (shown in Fig. 3 as clone 8) which showed both striking homology to the amino-terminal exon of other PLA_2 enzymes[45], as well as conservation of many of the known or proposed residues involved in catalysis. In particular, residues corresponding to the "Ca^{++}-binding loop" (Tyr_{28} through Gly_{32}), Phe_5, the amino-terminal amphipathic helix containing one positively-charged face (Ser_1 to Arg_{16}), and the conserved portion of "helix C" between the Ca^{++}-binding loop and the active site (Leu_{31} to Arg_{43}), are contained within this sequence. On the other hand, the substitution of Arg for Met_8, Val for Ile_9, and Asp for Gly_{33}, constitute changes in key residues not seen elsewhere in active PLA_2s[14,15,39]. A substitution of Arg_8 would likely perturb the amino-terminal helix and might interfere with folding, since Met_8 is located in the hydrophobic core of the protein (see Ref. 4). The Val_9 substitution would appear less dramatic; however Ile in this important position is 100% conserved in other PLA_2s. Studies elsewhere[64] have shown that several modifications of Ile_9 in porcine pancreatic PLA_2 resulted in reduced PLA_2 activity. Regarding the substitution of Asp for Gly_{33}, this residue in other enzymes constitutes one of the invariant glycine residues of the "Ca^{++}-binding loop". In the venom PLA_2 homolog of the Australian tiger snake Notechis scutatus, the substitution of a nearby Ca^{++}-binding loop glycine (residue 30) may result in a complete loss of PLA_2 activity. While these substitutions argue that clone 8 may encode an inactive PLA_2, it is also possible that substitutions elsewhere in the molecule could compensate for these alterations.

A resolution of this possibility may lie within the sequence of the other coding exons of the molecule which, in spite of the strong conservation within this particular exon of the gene, were either too divergent or too distant in chromosomal location to be detected by cross-hybridization to pancreatic PLA_2 cDNA.

```
Type                    1        10        20        30        40
I     VLLTVAAA DSGISPR AVWQFRKMIKCVIPGSDPFLEYNNYGCYCGLGGSGTPVDELDK
IIA   GLLQAHG  ------- NLVNFHRMIK-LTTGKEAALSYGFYGCHCGVGGRGSPKDATDR
8     VVAAPTHS ------- SFWQFQRRVK-HITGRSAFFSYYGYGCYCGLGDKGIPVDDTDR
10    VPAVQG   ------- GLLDLKSMIE-KVTGKNALTNYGFYGCYCGWGGRGTPKDGTDW
```

Figure 3. Human PLA_2 amino acid alignment. The duduced amino acids for one exon of human Type I (pancreatic) (Refs. 22 and 23), Type IIA (Refs. 34-38) (referred to elsewhere as RASF-A), clone 8 (referred to elsewhere [Ref. 45] as np PLA_2), and clone 10 (Ref. 47) were mutually aligned.

In a similar manner, another homologous putative PLA_2 coding exon was found in the process of obtaining the human gene for Type IIA PLA_2. Of the 13 clones obtained when screening a human genomic library with a 45-base oligonucleotide probe[36,47], three encoded an amino acid sequence which was different from any sequence previously observed. This sequence, shown in Fig. 3 and referred to as clone 10, also contained residues corresponding to most of the known PLA_2 structural features, including a "Ca^{+2}-binding loop", as well as conserved residues such as Met_8, Ile_9, Tyr_{22}, and Gly_{33}. The sequence also contains two key substitutions, however. The first, Leu for Phe_5, is an unusual one among PLA_2 sequences, seen only in the "K-49" enzyme from Agkistrodon venom[50] which has a very low specific activity[39]. The second is the loss of a positively charged residue at position 43 (Trp). The rat Type IIA PLA_2 sequence contains Glu at this position, and horse pancreatic PLA_2, the most active of the known pancreatic enzymes, contains Ala at this position[39].

At the DNA level, both of these sequences contain perfectly conserved intron/exon consensus sequences[49] at both ends, supporting their possible role as exons of a larger transcribed gene. In the case of clone 8, homologous sequences from porcine and rat DNA were cloned and sequenced[45]. The porcine and rat homologs encoded a strikingly conserved amino acid sequence (86% and 90% conservation, respectively). In order to more extensively examine the occurrence of both putative genes in other species, we examined their cross-species hybridization in more detail. As with Type IIA PLA_2 above, hybridization probes containing DNA coding regions from either clone 8 or clone 10 and including approximately

400 bp of flanking DNA were prepared by PCR. These probes were hybridized to identical blots of genomic DNA from the same species under similar hybridization stringencies. The results are shown in Fig. 2, panels B and C. For clone 8, single faintly hybridizing bands were observed in each mammalian species except for the sample from monkey which showed two bands. For clone 10, the probe apparently contained some repetitive sequences, since substantial background hybridization was observed, in several of the lanes and particularly in the lane containing human DNA. Nevertheless, single strongly hybridizing bands were observed especially in rat, mouse, monkey and canine DNA. It is thus evident that homologs of both of these gene elements are present in the genomes of most if not all of the mammals tested.

One or both of these sequences could encode a portion of an undescribed mammalian Ca^{++}-dependent PLA_2 enzyme. Since both sequences define a cysteine pattern of the Type II class, they could be considered Types IIB and IIC PLA_2, respectively. If this were the case, the remaining exons encoding the other portions of the molecule appear to be more divergent in DNA sequence than the exons discovered by cross-hybridization. Since the Ca^{++}-binding loop is, by far, the most conserved region of all known PLA_2 sequences, it is not surprising that the exon encoding it would be the easiest to detect by cross-hybridization. Another possibility is that these genes could encode a portion of another molecule with different enzymatic properties, which has obtained its Ca^{++}-binding domain by partial gene duplication of this portion of the PLA_2 sequence. In this regard, however, the Ca^{++}-binding loop of these putative PLA_2s bears no obvious resemblance or homology to the Ca^{++}-binding portions of other well-known molecules such as the endonexins and calpactins[33,34], or the troponin C superfamily members calmodulin, myosin or troponin[36,37]. A third, less interesting possibility is that these putative gene portions are functionless gene relics (i.e. pseudogenes). While this latter possibility remains open, it seems unlikely in light of the strikingly high degree of conservation of this exon sequence exhibited between its mammalian homologs. It is difficult to imagine such strong conservation of primary sequence without strong positive selective pressure imposed by a requirement for function.

A fourth possibility is that the two homologous exons are transcribed, but are spliced alternately to the other exons contained in the Type IIA PLA_2 gene. If this were the case, the alternate exon would likely reside in a chromosomal location near the rest of the gene. However, it would appear at the outset that both of these exons lie a significant distance from either of the two complete genes, since none of the genomic clones (15 kB average insert size) hybridized to either of the other cloned PLA_2 genes. Nevertheless, as a preliminary attempt to explore this possibility, we determined the chromosomal locations of the Types I and IIA PLA_2

genes as well as the two partial putative genes (clones 8 and 10) described above.

Chromosomal Localization of the PLA$_2$ Genes

DNA probes were prepared using either cDNAs for the two complete PLA$_2$ genes, or restriction fragments containing the two additional exons without repetitive sequences for the two partial putative genes. The DNA probes were hybridized to blots of predigested genomic DNA obtained from a panel of mouse/human somatic cell hybrids under stringent conditions[45]. Distinct bands corresponding to each of the human genes was observed, and were tabulated for the occurrence of the bands in each hybrid cell line. When this was done with Type I PLA$_2$, complete concordance was shown with chromosome 12[45]. Similar analyses in the same study showed that clone 8 mapped to chromosome 1[45]. Here, we have performed similar studies using probes for Types IIA PLA$_2$ and clone 10. The results, summarized in Table 1, suggest that these genes also reside on chromosome 1.

The chromosomal co-localization of the two homologous exons with Type IIA PLA$_2$ leaves open the possibility that the three genetic loci are related with respect to origin, and that they could indeed be alternate exons to the Type IIA PLA$_2$ gene. In this regard it is noteworthy that clone 8, which was detected with Type I PLA$_2$ cDNA probes, is actually closer in both sequence homology and gene location to the PLA$_2$ IIA gene. It should be possible to elucidate the role(s) of clones 8 and 10 through the determination of the distances separating the genes and their relative orientations. A full understanding of the role these gene homologs play will require the detection of transcription of the sequences in some tissue or cell type.

Other PLA$_2$s

Over the past two decades, there have been numerous reports of PLA$_2$ isolates from nearly every mammalian tissue or extract. Many of these reports described PLA$_2$ enzymes with dramatically different properties[55-61] and molecular weights[55-58]. Many may eventually prove to be identical to one of the forms characterized here, and others may eventually be shown to constitute distinct enzyme forms. There is at least one other PLA$_2$ which is Ca^{++}-independent and has been characterized in the cytosol fraction from several rat tissues[62].

One PLA2 isolate which has demonstrated distinct biochemical properties from the two known enzymes was isolated as a second peak of activity (peak B) present in synovial fluid[35]. This

Table 1. Human chromosomal mapping of Type IIA PLA2 and clone 10 genes. DNA probes made from Type IIA cDNA and gene coding regions from clone 10 were used to probe blots of genomic DNA from a panel of somatic cell hybrids derived from the fusion of thymidine kinase-deficient mouse cells and normal human male fibroblasts (IMR91). The columns, left to right, indicate the hybrid name, the hybridization results for both probes, and whether or not each human chromosome is present in the hybrid. The number of discordant hybrids (wrong hybridization bands relative to each chromosome) is shown in the bottom row.

HUMAN CHROMOSOMES

Hybrid Clone	Clone 10*	Type IIA PLA2*	1	2	3	4	5	6	7	8	9	10	11	12	13	14	15	16	17	18	19	20	21	22	X	Y
84-2	+	+	+	+	+	+	+	+	+	+	-	-	+	+	-	+	+	+	+	+	+	+	+	+	-	-
84-3	+	+	+	+	+	+	+	+	+	+	-	+	-	-	+	+	+	+	+	+	+	+	-	-	-	-
84-4	+	+	+	+	+	+	+	+	+	+	-	+	-	-	+	+	+	+	+	+	+	-	+	-	-	-
84-5	-	-	-	-	-	-	+	+	+	+	-	+	+	+	(+)	+	+	+	-	-	-	+	+	+	+	+
84-7	-	-	-	+	+	+	+	+	+	+	-	-	+	+	-	-	+	-	+	-	-	-	+	+	-	-
84-13	-	-	-	+	+	-	+	-	-	-	-	+	+	+	+	+	-	-	+	+	-	-	+	+	-	-
84-20	-	(+)	+	+	-	+	-	-	-	-	(+)	(+)	(+)	+	-	-	-	+	+	-	+	(+)	(+)	(+)	(+)
84-21	+	+	+	-	-	+	+	+	+	+	-	+	-	+	+	+	+	-	+	-	+	+	+	-	-	+
84-25	-	-	-	-	-	+	+	+	+	+	-	+	+	+	-	+	-	-	+	-	+	+	+	-	-	+
84-26	+	+	+	-	+	+	+	+	+	+	-	+	-	(+)	+	+	+	+	+	-	(+)	+	+	+	+
84-27	-	-	-	-	-	-	(+)	-	-	-	-	-	-	+	-	-	-	+	+	-	(+)	+	-	-	-
84-30	-	-	-	-	+	-	+	-	-	-	+	-	(+)	-	-	-	-	+	-	(+)	(+)	-	-	-	+
84-34	-	-	-	-	+	+	-	+	-	-	-	-	-	-	+	-	+	+	(+)	(+)	+	+	-	-	-	+
84-35	-	-	-	-	-	+	-	+	+	+	-	-	+	+	-	-	+	-	+	+	-	-	+	+	-	-
84-37	-	-	+	-	+	+	+	+	+	+	+	+	+	+	+	+	+	+	+	+	+	+	+	+	+	+
84-38	+	+	+	-	-	+	(+)	+	+	+	+	+	-	+	+	-	+	-	+	+	+	(+)	+	+	+	-
84-39	-	-	-	+	+	+	-	-	+	+	-	+	-	+	-	-	+	-	+	-	-	-	-	-	-	-
Number of discordant hybrids			0	5	9	8	8	10	6	9	6	5	10	7	4	7	6	5	11	6	2	6	11	8	7	5

* "+" indicates PLA2 sequences in the hybrid clone determined by the presence of the human band; "-" indicates absence of the gene.

** "+" indicates presence of the human chromosome in greater than 30% of metaphases analyzed; "(+)" indicates presence of the chromosome in 10-30% of the metaphases analyed; "-" indicates absence of the human chromosome.

enzyme showed a strikingly different profile of substrate preference, sensitivity to ionic strength and detergent concentration, and chromatographic behavior when assayed with mixed micelles, as compared to the Type IIA PLA_2 enzyme present in the same fluid. Preliminary amino-terminal amino acid sequencing data suggested the sequence was distinct from any of the known PLA_2 forms, however the data has so far been inconclusive.

CONCLUSIONS

The use of improved isolation techniques, amino acid sequencing and cDNA cloning has in the past 3 years clarified the relationships among many of the Ca^{++}-dependent PLA_2 forms expressed in different tissues. Currently, two complete genes encoding Ca^{++}-dependent phospholipases A_2 have been isolated, sequenced, and mapped to their respective chromosomal locations. The Type I PLA_2 gene shows a relatively limited pattern of expression in specialized tissues. In contrast, the Type IIA PLA_2 gene product appears to be expressed by a wide variety of tissue and cell types and accumulates in certain body fluids associated with inflammatory reactions. These observations suggest that the Type IIA PLA_2 gene product is present in a diverse array of inflammatory conditions and may contribute in a major way to the generation of eicosanoid products. It is now clear that homologous forms of this gene exist and are transcribed in other mammals. These results may have significance when considering animal models in which to test putative PLA_2-inhibiting therapeutics for use in human diseases. In addition, we have demonstrated chromosomal linkage of the Type IIA PLA_2 gene with two other related genetic elements which display significant homology to one exon of the Type IIA PLA_2 gene. While the exact roles of and relationships between these three loci need to be determined in more detail, it is possible that these additional exons could be utilized to provide a Type II PLA_2 gene product with further diversity in form, and therefore function.

ACKNOWLEDGEMENTS

This work was supported by NIH grant HL28481 (to AJL). AJL is an Established Investigator of the American Heart Association. We thank Dr. T. Mohandas, Harbor-UCLA Medical Center, for making available the mouse-human somatic cell hybrid panel. We also thank Dr. H. Verheij for constructive comments and criticism and Kayo Miyasaki for preparation of the manuscript.

REFERENCES

1. L.M.G. Van Golde, and S.G. van den Bergh, Introduction: general pathway in the metabolism of lipids in mammalian tissues, in: "Lipid Metabolism in Mammals", F. Snyder, ed., Plenum Publishing Corp., New York (1977).

2. P. Elsbach, J. Weiss, R. Franson, S. Beckerdite-Quagliata, A. Schneider, and L. Harris, Separation and purification of a potent bactericidal/permeability-increasing protein and a closely associated phospholipase A_2 from rabbit polymorphonuclear leukocytes, J. Biol. Chem. 254:11000 (1979).

3. M.F. Heath, and W. Jacobson, The action of lung lysosomal phospholipases on dipalmitoyl phosphotidylcholine and its significance for the synthesis of pulmonary surfactant, Pediatr. Res. 14:254 (1980).

4. H.M. Verheij, A.J. Slotboom, and G.H. De Haas, Structure and function of phospholipase A_2, Rev. Physiol. Biochem. Pharmacol. 91:91 (1981).

5. R.F. Irvine, How is the level of free arachidonic acid controlled in mammalian cells?, Biochem. J. 204:3 (1982).

6. F. Snyder, Chemical and biochemical aspects of platelet activating factor: a novel class of acetylated ether-linked choline-phospholipids, Med. Res. Rev. 5:107 (1985).

7. P. Vadas and W. Pruzanski, Role of Extracellular Phospholipase A2 in Inflammation, in: "Adv. in Inflamm. Res.", I. Otterness, R. Capetola, and S. Wong, eds., Raven Press, New York 7:51 (1984).

8. W. Pruzanski, P. Vadas, J. Kim, H. Jacobs, and E. Stefanski, Phospholipase A2 activity associated with synovial fluid cells, J. Rheumatol. 15:791 (1989).

9. P. Vadas, W. Pruzanski, E. Stefanski, L. Johnson, J. Seilhamer, R. Mustard, Jr., and J. Bohnen, Phospholipase A2 in acute bacterial peritonitis in man, in: "Cell Activation and Signal Initiation: Receptor and Phospholipase Control of Inositol Phosphate, PAF, and Eicosanoid Production," E. Dennis, ed., Alan R. Liss, New York (1989).

10. P. Vadas, W. Pruzanski, and E. Stefanski, Extracellular phospholipase A2: causative agent in circulatory collapse of septic shock?, Agents and Actions 24:320 (1988).

11. P. Vadas, W. Pruzanski, E. Stefanski, B. Sternby, R. Mustard, J. Bohnen, I. Fraser, V. Farewell, and C. Bombardier, Pathogenesis of hypotension in septic shock: correlation of circulating phospholipase A2 levels with circulatory collapse, Crit. Care Med. 16:1 (1988).

12. P. Vadas, W. Pruzanski, E. Stefanski, J. Ruse, V. Farewell, J. McLaughlin, and C. Bombardier, Concordance of endogenous cortisol and phospholipase A2 levels in gram-negative septic shock: a prospective study, J. Lab Clin. Med. 111:584 (1988).

13. P. Vadas, W. Pruzanski, J. Kim, and V. Fornasier, The proinflammatory effect of intra-articular injection of soluble

human and venom phospholipase A2, $\underline{Am}$. $\underline{J}$. $\underline{Pathol}$. 134:807 (1989).

14. M.J. Dufton, D. Eaker and R.C. Hider, Concormational properties of phospholipases A2: Secondary structure prediction, circular dichroism and relative interface hydrophobicity, $\underline{Eur}$. $\underline{J}$. $\underline{Biochem}$. 137:537 (1983).

15. M.J. Dufton and $\underline{R.C}$. $\underline{Hider}$, Classification of phospholipase A_2 according to sequence. Evolutionary and pharmacological implications, $\underline{Eur}$. $\underline{J}$. $\underline{Biochem}$. 137:545 (1983).

16. R. Renetseder, S. Brunie, B.W. Dijkstra, J. Drenth, and P.B. Sigler, A comparison of the crystal structures of phospholipase A_2 from bovine pancreas and Crotalus atrox venom, $\underline{J}$. $\underline{Biol}$. $\underline{Chem}$. 260:11627 (1985).

17. A.L. Slotboom, H.M. Verheij, and G.H. De Haas, On the mechanism of phospholipase A_2, $\underline{New}$ $\underline{Comp}$. $\underline{Biochem}$. 4:354 (1982).

18. R.L. Heinrikson, E.T. Krueger, and P.S. Keim, $\underline{J}$. $\underline{Biol}$. $\underline{Chem}$. 252:4913 (1977).

19. R.M. Kini, and H.J. Evans, Structure-function relationships of phospholipases: the anticoagulant region of phospholipase A2, $\underline{J}$. $\underline{Biol}$. $\underline{Chem}$. 262:14402 (1987).

20. W.C. Puijk, H.M. Verheij, and G.H. De Haas, The primary structure of phospholipase A2 from porcine pancreas: a reinvestigation. $\underline{Biochim}$. $\underline{Biophys}$. $\underline{Acta}$ 492:254 (1977).

21. O. Ohara, M. Tamiki, E. Nakamura, Y. Tsuruta, Y. Fujii, M. Shin, H. Teraoka, and M. Okamoto, Dog and rat pancreatic phospholipases A2: Complete amino acid sequences deduced from complimentary DNAs. $\underline{J}$. $\underline{Biochem}$. 99:733 (1986).

22. H.M. Verheij, J. Westerman, B. Sternby, and G. De Haas, The complete primary sequence of phospholipase A2 from human pancreas, $\underline{Biochim}$. $\underline{Biophys}$. $\underline{Acta}$ 747:93 (1983).

23. J.J. Seilhamer, T.L. Randall, M. Yamanaka, and L.K. Johnson, Pancreatic phospholipase A_2: Isolation of the human gene and cDNAs from porcine pancreas and human lung, $\underline{DNA}$ 5:519 (1986).

24. R. Verger, F. Ferrato, C.M. Mansback, and G. Pieroni, Novel intestinal phospholipase A2: purification and some molecular characteristics. $\underline{Biochemistry}$ 21:6883 (1982).

25. H.W. Chang, I. Kudo, M. Tomita, and K. Inoue, Purification and characterization of extracellular phospholipase A2 from peritoneal cavity of caseinate-treated rat, $\underline{J}$. $\underline{Biochem}$. 102:147 (1987).

26. M. Hayakawa, I. Kudo, M. Tomita, and K. Inoue, Purification and characterization of a membrane-bound phospholipase A2 from rat platelets, $\underline{J}$. $\underline{Biochem}$. 103:263 (1988).

27. M. Hayakawa, I. Kudo, M. Tomita, S. Nojima, and K. Inoue, The primary structure of rat platelet phospholipase A_2, $\underline{J}$. $\underline{Biochem}$. 104:767 (1988).

28. T. Ono, H. Tojo, S. Kuramitsu, H. Kagamiyama, and M. Okamoto, Purification and characterization of a membrane-associated phospholipase A_2 from rat spleen: its comparison with a

cytosolic phospholipase A_2 S-1, J. Biol. Chem. 263:5732
(1988).

29. M. Hayakawa, K. Horigome, I. Kudo, M. Tomita, S. Nojima, and
    K. Inoue, Amino acid composition and NH_2-terminal amino acid
    sequence of rat platelet secretory phospholipase A_2, J.
    Biochem. 101:1311 (1987).

30. A.J. Aarsman, J.G. de John, E. Arnoldussen, F.W. Neys, P.D.
    van Wassenaar, and H. Van den Bosch, Immunoaffinity
    purification, partial sequence, and subcellular localization
    of rat liver phospholipase A_2, J. Biol. Chem. 264:10008
    (1989).

31. J. Forst, J. Weiss, P. Elsbach, J.M. Maranganore, I. Reardon,
    and R.L. Heinrikson, Biochemistry 25:8381 (1986).

32. C. E. Ooi, G. Wright, J. Weiss, and P. Elsbach, Purification
    to homogeneity and properties of rabbit granulocyte PLA_2,
    Clin. Res. 36:465A (1988).

33. H. Mizushima, I. Kudo, K. Horigome, M. Murakami, M. Hayakawa,
    D.K. Kim, E. Kondo, M. Tomita, and K. Inoue, Purification of
    rabbit platelet secretory phospholipase A_2 and its
    characteristics, J. Biochem. 105:520 (1989).

34. S. Hara, I. Kudo, K. Matsuta, T. Miyamoto, and K. Inoue, Amino
    acid composition and NH_2-terminal amino acid sequence of human
    phospholipase A_2 purified from rheumatoid synovial fluid, J.
    Biochem. 104:326 (1988).

35. J. Seilhamer, S. Plant, W. Pruzanski, J. Schilling, E.
    Stefanski, P. Vadas, and L. Johnson, Multiple forms of
    phospholipase A_2 in arthritic synovial fluid, J. Biochem.
    106:730 (1989).

36. J.J. Seilhamer, P. Vadas, S. Plant, J.A. Miller. J. Kloss, W.
    Pruzanski, and L.K. Johnson, Cloning and reconbinant
    expression of phospholipase A_2 present in rheumatoid arthritic
    synovial fluid, J. Biol. Chem. 264:5335 (1989).

37. R.M. Kramer, C. Hession, B. Johansen, G. Hayes, P. McGray,
    E.P. Chow, R. Tizzard, and R.B. Pepinski, Structure and
    properties of a human non-pancreatic phospholipase A_2, J.
    Biol. Chem. 264:5768 (1989).

38. C. Lai and K. Wada, Phospholipase A_2 from human synovial
    fluid:purification and structural homology to the placental
    enzyme, Biochem. Biophys. Res. Commun. 157:488 (1988).

39. C. van den Bergh, A. Slotboom, H. Verheij, and G. de Haas, The
    role of Asp-49 and other conserved amino acids in
    phospholipases A_2 and their importance for enzymatic activity,
    J. Cell. Biochem. 39:379 (1989).

40. H. Tojo. T. Ono, S. Kuramitsu, H. Kagamiyama, and M. Okomoto,
    A phospholipase A_2 in the supernatant fraction of rat spleen:
    its similarity to rat pancreatic phospholipase A2, J. Biol.
    Chem. 263:5724 (1988).

41. H. Tojo, T. Ono, and M. Okamoto, A pancreatic-type
    phospholipase A_2 in rat gastric mucosa, Biochem. Biophys. Res.
    Commun. 151:1188 (1988).

42. Y. Matsuda, M. Ogawa, T. Shibata, K. Nakaguchi, J. Nishijima, C. Wakasugi, and T. Mori, Distribution of immunoreactive pancreatic phospholipase A_2 (IPPL-2) in various tissues, Res. Commun. in Chem. Pathol. and Pharmacol. 58:281 (1987).

43. D. Bar-Sagi, J. Suhan, F. McCirmick, and J. Feramisco, Localization of phospholipase A_2 in normal and ras-transformed cells, J. Cell Biol. 106:1649 (1988).

44. T. Nevalainen, J. Escola, A. Aho, V. Havia, T. Lovgren, and V. Nanto, Immunoreactive phospholipase A_2 in serum in acute pancreatitis and pancreatic cancer, Clin. Chem. 31:1116 (1985).

45. J.J. Seilhamer, T.L. Randall, L.K. Johnson, C. Heinzmann, I. Klisak, R.S. Sparkes, and A.J. Lusis, Novel gene exon homologous to pancreatic phospholipase A_2: sequence and chromosomal mapping of both human genes, J. Cell. Biochem. 39:327 (1988).

46. P. Lind and D. Eaker, Complete amino-acid sequence of a non-neurotoxic, non-enzymatic phospholipase A_2 homolog from the venom of the Australian tiger snake Notechis scutatus, Eur. J. Biochem. 111:403 (1980).

47. J. Seilhamer, S. Frank, P. Vadas, W. Pruzanski, A.J. Lusis, and L.K. Johnson, Chromosomal mapping of human synovial fluid PLA_2 and a related exon, manuscript in preparation.

48. M.A. Innis, K.B. Myambo, D.H. Gelfand, and M.D. Brow, DNA sequencing with Thermus aquaticus DNA polymerase and direct sequencing of polymerase chain reaction-amplified DNA, Biochemistry 85:9436 (1988).

49. S. Mount, A catalogue of splice junction sequences. Nucleic Acids Res. 10:459 (1982).

50. J. Maranganore and R. Heindrikson, The Lysine-49 phospholipase A_2 from the venom of Agkistrodon piscivorus piscivorus: relation of structure and function to other phospholipases A_2, J. Biol. Chem. 261:4797 (1986).

51. M.J. Geisow, Common domain structure of Ca^{2+} and lipid-binding proteins, Febs. Lett. 203:99 (1986).

52. K. Weber, and N. Johnsson, Repeating sequence homologies in the p36 target protein of retroviral protein kinases and lipocortin, the p37 inhibitor of phospholipase A_2, Febs. Lett. 203:95 (1986).

53. M.O. Dayhoff, Atlas of Protein Sequence and Structure, Suppl. 5:273 (1978).

54. R.H. Kretsinger, Structure and evolution of calcium-modulated proteins, CRC Crit. Rev. Biochem. 8:119 (1980).

55. S. Sahu, and W.S. Lynn, Characterization of phospholipase A from pulmonary secretions of patients with alveolar proteinosis. Biochim. Biophys. Acta 489:307 (1977).

56. N.C.C. Gray, and K.P. Strickland, The purification and characterization of a phospholipase A_2 activity from the 106000 X G pellet (microsomal fraction) of bovine brain acting on phosphatidylinositol. Can. J. Biochem. 60:108 (1982).

57. R.M. Kramer, C. Wuthrich, C. Bollier, P.R. Allegrini, and P. Zahler, Isolation of phospholipase A_2 from sheep erythrocyte membranes in the presence of detergents, Biochim. Biophys. Acta 507:381 (1978).

58. L. Aron, S. Jones, and C. J. Fielding, Human plasma lecithin-cholesterol acyl transferase, J. Biol. Chem. 253:7220 (1978).

59. R. Ulevitch, Y. Watanabe, M. Sano, M. Lister, R. Deems, and E. Dennis, Solubilization, purification, and characterization of a membrane-bound phospholipase A_2 from the p388D1 macrophage-like cell line, J. Biol. Chem. 263:3079 (1988).

60. P. Antaki, J. Langais, P. Ross, P. Guerette, and K. Roberts, Evidence for two forms of phospholipase A_2 in human semen, Gamete Research 18:305 (1988).

61. S. Parthasarathy, U. Steinbrecher, J. Barnett, J. Witzum, and D. Steinberg, Essential role of phospholipase A_2 activity in endothelial cell cell-induced modification of low density lipoprotein, Proc. Nat. Acad. Sci. U. S. A. 82:3000 (1985).

62. A.J. Pierik, J. Nijssen, A. Aarsman, H. Van den Bosch, Calcium-independent phospholipase A_2 in rat tissue cytosols. Biochim. Biophys. Acta 962:345 (1988).

63. J. Seilhamer, P. Vadas, W. Pruzanski, S. Plant, E. Stefanski, and L. Johnson, Synovial fluid phospholipase A_2 in arthritis, in: "Therapeutic Approaches to Inflammatory Diseases", A.J. Lewis, N.S. Doherty, and N.R. Ackerman, ed., Elsevier, New York (1989).

64. van Binsbergen, et. al., manuscript in preparation.

STRUCTURE AND PROPERTIES OF A SECRETABLE PHOSPHOLIPASE A$_2$ FROM HUMAN PLATELETS

Ruth M. Kramer[a], Berit Johansen[b], Catherine Hession[c] and R. Blake Pepinsky[c]

[a] Lilly Research Laboratories, Indianapolis, IN; [b] Norwegian Institute of Technology, Trondheim, Norway; [c] Biogen Inc., Cambridge, MA

Intracellular cytosolic and secretable phospholipases A$_2$

Phospholipases A$_2$ (PLA$_2$s) constitute a diverse family of enzymes that hydrolyze the sn-2 fatty acyl ester bond of phosphoglycerides liberating free fatty acids and lysophospholipids (Dennis, 1983). Mammalian extracellular PLA$_2$s are abundant in pancreatic secretions, but are also present in plasma, lymph and pulmonary alveolar secretions (Vadas and Pruzanski, 1986). Intracellular PLA$_2$s are found in all tissues and cells (van den Bosch, 1980), where they are located either in the cytosolic compartment associated with the plasma membrane or stored within organelles of the vacuolar system, such as secretory granules and lysosomes. The granule-associated PLA$_2$s are designed to effectively degrade phospholipids upon exocytosis thereby serving either a digestive (Verheij et al., 1981) or an anti-microbial function (Elsbach, 1980). In contrast, the cytosolic PLA$_2$s play a key role in the metabolism of cellular lipids, including the biosynthesis of specifically tailored phospholipids and the degradation of peroxidized phospholipids thus protecting membranes from oxidation damage (Waite, 1987). Furthermore, such PLA$_2$s are involved in the generation of rate-limiting precursors for various types of lipid mediators (Irvine, 1982; Snyder, 1985) that transmit stimulatory signals to other cells or function as intracellular messengers (O'Flaherty, 1982; Larsen and Henson, 1983). This amplification mechanism is an integral part of the inflammatory response of tissues to injury that normally leads to removal of the inciting agent and repair of injured site.

The control and regulation of enzymatic activity of the granule-associated and cytosolic PLA_2 are quite distinct. PLA_2s sequestered in granules under conditions that suppress catalytic activity are activated upon secretion into the extracellular space, where conditions for enzyme activity are favorable and/or conversion of proenzyme to fully active enzyme occurs (Verheij et al., 1981). The regulation of the activity of cytosolic PLA_2s, on the other hand, appears to be subject to many different factors that may act in concert to modulate catalytic activity and direct the enzyme to appropriate target substrates. Thus, changes in intracellular free Ca^{2+} (van den Bosch, 1980), perturbations of the plasma membrane (Dawson et al., 1983), protein kinase C (Waite, 1987) and PLA_2-activating proteins antigenically related to melittin (Clark et al., 1987) have all been implicated in PLA_2 activation. More recently, it has been proposed that cytosolic PLA_2s may be directly activated via membrane receptors and guanine nucleotide-binding proteins (Axelrod et al., 1988).

Human platelet secretable phospholipase A_2

Extracellular, secreted PLA_2s from snake venoms and pancreas have been sequenced and structurally defined, and the molecular mechanism of enzymatic action has been extensively studied (Verheij et al., 1981; Dennis, 1983). In contrast, until recently, little was known about the molecular structure and properties of non-pancreatic mammalian PLA_2s. Knowledge of the structure and mode of action of these PLA_2s is a prerequisite for understanding their involvement in cellular processes and elucidation of their role in disease.

We have previously characterized a PLA_2 from human platelets that exhibits functional properties typical for a cytosolic PLA_2 (Kramer et al., 1986, 1987 and 1988). The enzyme is activated by physiological (submicromolar) concentrations of Ca^{2+}, stimulated by diacylglycerol, hydrolyzes both diacyl-phospholipids and ether-linked PAF precursor phospholipids, and is catalytically active at neutral pH. This PLA_2 readily hydrolyzes phosphatidylcholine presented in the physical form of sonicated liposomes, has an apparent molecular weight of 60,000 (by gel filtration) and is a rather labile enzyme that is inactivated by most detergents, acid and nonphysiological salt concentrations. The scarceness of the enzyme in platelets and its lability have hampered purification of the enzyme.

We observed a second PLA_2 resembling an extracellular enzyme that is secreted by platelets upon stimulation with thrombin and collagen (Kramer et al., 1989). It has a similar pH-optimum as the cytosolic PLA_2, but its Ca^{2+}-dependence is in the millimolar range. Moreover, it shows a striking preference for substrate presented in the form of *E.coli* membranes and hydrolyzes phosphatidylcholine presented as sonicated liposomes at rates that are 750 times slower. Furthermore, this PLA_2 exhibits a remarkable stability to detergents and low pH (<1). Although the secretable PLA_2 is only a trace protein component of platelets, its great stability in acid was a key property that ultimately allowed us to purify the enzyme to homogeneity (Kramer et al., 1989). The results of that purification are summarized in Table 1 and Fig. 1. Using the solubilization procedure developed for an acid-stable PLA_2 from rabbit neutrophils (Elsbach et al., 1979), the PLA_2 was extracted from lysed platelets at pH 1. Approximately 90% of the extracted platelet protein was precipitated upon dialysis of the acid-extract against acetate buffer (pH 4.5), while the total enzymatic activity in the soluble fraction increased by 39-fold. The dialysate was clarified by centrifugation and the enzyme purified through sequential chromatography steps including cation exchange chromatography on S-Sepharose (Fig. 2A), gel filtration on Sephadex G-50 (Fig. 2B) and reverse-phase HPLC on a C_4-column (Fig. 2C). From HPLC the PLA_2 was recovered as a single peak that eluted with 35% acetonitrile. Overall, the platelet PLA_2 was purified over a million-fold with a recovery of 34%. The purified platelet PLA_2 had an apparent molecular weight of 13,000 (by gel filtration), exhibited maximal activity at pH 8-9 in the presence of 10 mM Ca^{2+} (Fig. 3A and B), showed a great preference for *E.coli* membrane substrate and in mixed micelles hydrolyzed phosphatidylethanolamine more readily than phosphatidylcholine (Kramer et al., 1989). For amino-terminal sequence analysis, platelet PLA_2 was subjected to SDS-PAGE and electroblotted onto Immobilon paper and stained with Coomassie Blue (Fig. 1). The band at 14,000 was excised and directly subjected to sequence analysis. The amino-terminal sequence was determined to be: NH_2-Asn-Leu-Val-Asn-Phe-His-Arg-Met-Ile-Lys-Leu-Thr-Thr-Gly-Lys-Glu-Ala-Ala-Leu.

Based on the sequences Asn^4 to Ile^9 and Met^8 to Thr^{13}, two overlapping degenerate oligonucleotide probes (17 nucleotides in length; 48- and 144-fold redundant, respectively) were synthesized and used to screen a

Table 1.   Purification of PLA$_2$ from human platelets

Purification step	Protein (mg)	PLA$_2$ activity (units*)	PLA$_2$ activity (units/mg)	Yield (%)	Purif. (-fold)**
(0) Sonication	7510	0.08	0.000011		
(1) Extraction (pH 1)	6060	0.13	0.000021		
(2) Dialysis (pH 4.5)	718	5.01	0.007	100	10
(3) S-Sepharose	2.5	2.07	0.83	41	1,186
(4) Sephadex G-50	0.1	4.81	48.1	96	68,714
(5) HPLC	~0.002	1.71	855	34	>1,221,143

*   units = μmol substrate (*E.coli* phosphatidylethanolamine + phosphatidylglycerol) hydrolyzed/15 min
**   adjusted for apparent increase in total PLA$_2$ activity

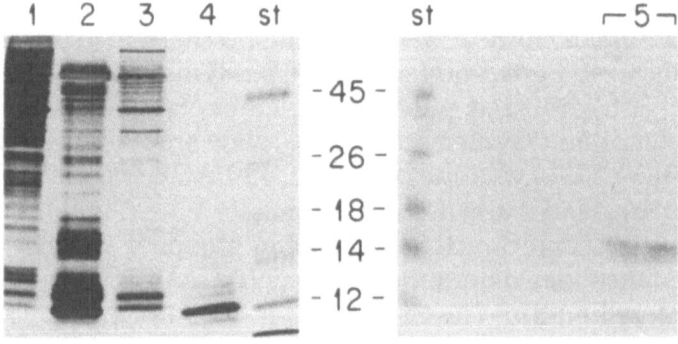

Fig. 1.   Purification of PLA$_2$ from human platelets by steps (1) - (5) as indicated above followed by SDS-PAGE with silver (1-4) and Coomassie (5) staining (st, standards).

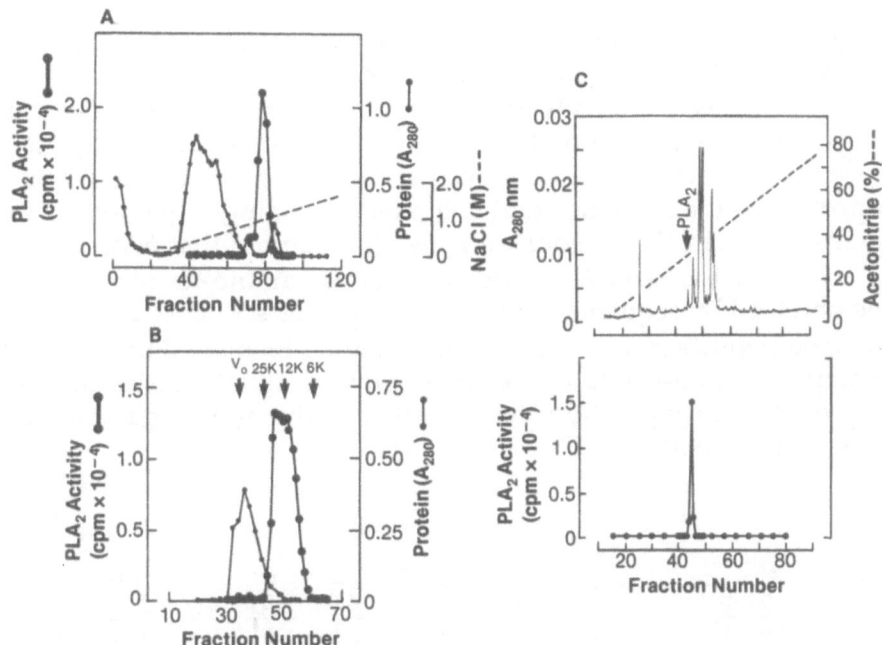

Fig. 2.   Elution profiles of protein and PLA$_2$ activity (assayed in aliquots of fractions using *E.coli* substrate) from S-Sepharose (A), Sephadex G-50 (B) and C$_4$-reverse-phase HPLC (C) columns.

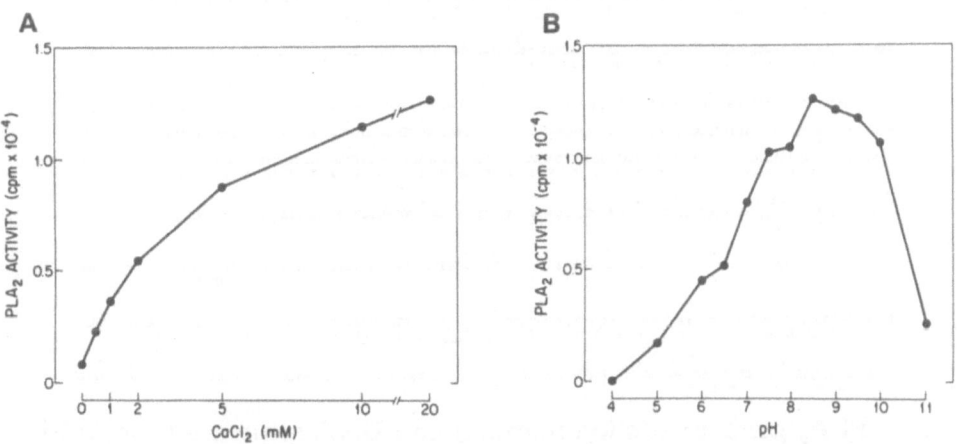

Fig. 3.   Ca^{2+} (A) and pH (B) dependence of human platelet PLA$_2$ activity.

human genomic phage library by plaque hybridization analysis. One clone hybridized to both probes and contained a genomic DNA insert of 15 kb. This insert comprised a 6.2-kb HindIII restriction fragment that hybridized to both probes. The restriction fragment was subcloned and completely sequenced from both strands. The restriction map of the 6.2-kb fragment and a portion of the sequence that contains the coding sequence for the PLA$_2$s are shown in Fig. 4. Four exons were identified that encode a 144-amino acid protein. Exon 2 encodes the mature amino-terminal sequence that we obtained by sequencing platelet PLA$_2$. This sequence is preceded by a 20-residue peptide starting with a methionine. Exon 4 contains an in-frame stop codon and putative polyadenylation signal AATAAA. The intron-exon structure greatly resembles that of the human pancreatic PLA$_2$ (Seilhamer et al., 1986).

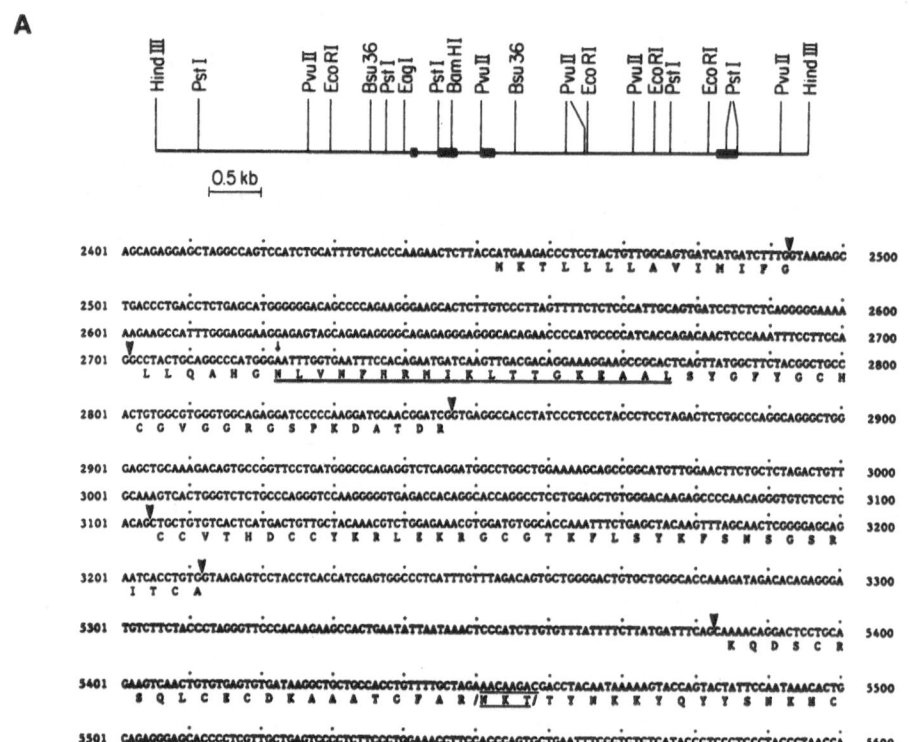

Fig. 4.   PLA$_2$ gene: restriction map (A) and DNA sequence (nucleotides 2401-3300 and 5301-5600) containing translated regions (B); triangles indicate intron-exon boundaries.

Southern blot analysis of human chromosomal DNA digests with a PLA$_2$ 1.4-kb Bsu36 restriction fragment (spanning nucleotides 2055-3414) revealed that a single gene exists for this human PLA$_2$ (Fig. 5A). A 3.8-kb PLA$_2$ restriction fragment was cloned into the vector pBG341 and the resulting plasmid transfected into COS-7 cells. Conditioned media from transfected cells contained >100 times more PLA$_2$ activity than cells transfected with the vector only. Expression of the PLA$_2$ gene in animal cells confirmed that the gene encodes a functional protein. Furthermore, the fact that transfected cells secrete PLA$_2$ indicates that the peptide encoded by nucleotides 2453-2492 and 2704-2721 serves as a functional signal sequence. Levels of mRNA in transfected COS-7 cells were monitored by Northern blotting, probing with the PLA$_2$ 1.4-kb Bsu36 restriction fragment. The probe hybridized to a ~1200-base nucleotide in cells transfected with the PLA$_2$ construct (Fig. 5B) consistent with the predicted size of a PLA$_2$ transcript.

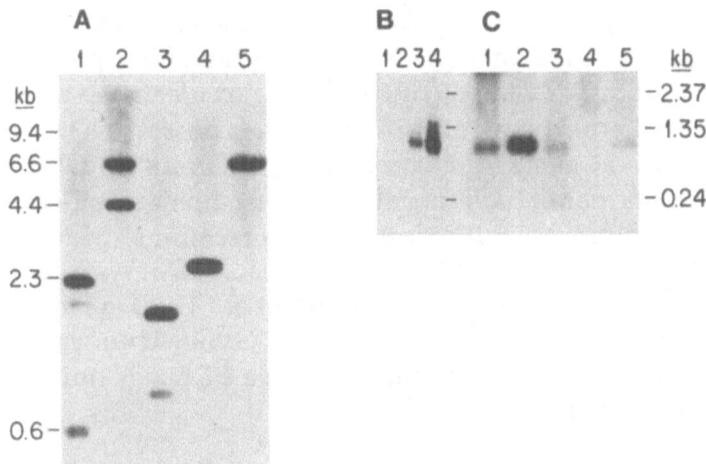

Fig. 5.   Southern analysis (A) of human genomic DNA digested with PstI (1), BamHI (2), PvuII (3), EcoRI (4) and HindIII (5) and probed with the Bsu36 restriction fragment; Northern blotting (B) of RNA from control (1,2) and PLA$_2$-transfected (3,4) COS-7 cells; Northern analysis (C) of RNA human tonsils (1), placenta (2), kidney (3) and synovial cells from patients with rheumatoid arthritis (4, 5).

Structural features of human platelet secretable phospholipase A_2

   Several important structural features of the human PLA_2 are predicted
from the amino acid sequence. First, the 20-amino acid extension
preceding the amino terminus of the mature protein resembles a signal
sequence for translocation from the cytosolic compartment to the central
vacuolar system (Blobel, 1980). In fact, the carboxyl-terminal amino acid
residues (-3, -1) of the peptide represent an acceptable signal peptidase
recognition site (Heijne, 1985). The presence of this element confirms that
the platelet PLA_2 is destined for the secretory pathway. Second, the
deduced amino acid sequence of the mature protein (Fig. 4) contains highly
conserved amino acid residues and sequences characteristic of all PLA_2s
sequenced to date (Slotboom et al., 1982; Heinrikson, 1982; Achari et al.,
1987): 1) the α-helical amino-terminal segment containing the lipophilic
residues Leu^2, Phe^5 and Ile^9; 2) the calcium binding loop with the typical
glycine-rich sequence Tyr^{25}-Gly-Cys-X-Cys-Gly-X-Gly-Gly-X-Gly-X-Pro^{37} and
the residue Asp^{49}; and 3) the active site residues His^{48}, Asp^{99}, Tyr^{52} and
Tyr^{73}. Two classes of PLA_2s have been proposed (Heinrikson et al., 1977):
Group I PLA_2s comprising PLA_2s from pancreatic juice and the venoms of
Elapidae (cobras) and Hydrophidae (sea snakes); and Group II PLA_2s com-
posed of PLA_2s from the venoms of Crotalidae (rattlesnakes and pit viper)
and Viperidae (old world vipers). Group I PLA_2s possess characteristic half
cystines at residues 11 and 77 that are missing in Group II PLA_2s. Group II
PLA_2s, on the other hand, have a unique half-cystine at residue 50 and an
extension of several amino acids at the amino terminus, which ends in a
half-cystine. Despite these structural differences X-ray crystallographic
analysis of bovine pancreatic PLA_2 (Dijkstra et al., 1978), a Group I enzyme
and *C.atrox* venom PLA_2 (Keith et al., 1981), a Group II enzyme, has shown
that the three-dimensional architecture of these PLA_2s is quite similar
(Renetseder et al., 1985).

   As shown in Fig. 6, the human PLA_2 sequence exhibits the half-cystine
pattern typical of Group II enzymes lacking both Cys^{11} and Cys^{77}, but
having Cys^{50} and the carboxyl-terminal extension ending with a half-
cystine. Overall, the homology with Group II enzymes is greater (44%)
than with Group I enzymes (37%). However, the human PLA_2 also
possesses a Group I-like feature within the structural segment identified as
a surface loop (residue 54-66) that is present in all pancreatic enzymes.
While in Group II PLA_2s eight residues are deleted in that segment, the
human PLA_2, like many Group I elapid snake venom PLA_2s, is missing

only six residues out of the total thirteen residues (Dufton et al., 1983).
Using site directed mutagenesis it has recently been demonstrated that
structural modifications within this surface loop greatly affect catalytic
activity and substrate preference of PLA$_2$ (Kuipers et al., 1989). The configu-
ration of the human platelet PLA$_2$ surface loop may thus, at least partly,
determine its distinctive substrate preference. The human platelet PLA$_2$ is
a cationic protein with an isoelectric point at pH > 10.5 containing structural segments enriched with positively charged residues. Notably, there is
a cluster of basic amino acids near the amino-terminus (Arg[7], Lys[10] and
Lys[15]). Such positively charged residues were proposed to be an important
determinant in the interaction of PLA$_2$s with specific biological targets,
namely native *E.coli* envelopes exposed to a bactericidal protein of neutrophils (Elsbach et al., 1986). Furthermore, the structural segment of the
human platelet PLA$_2$ corresponding to the surface loop contains three

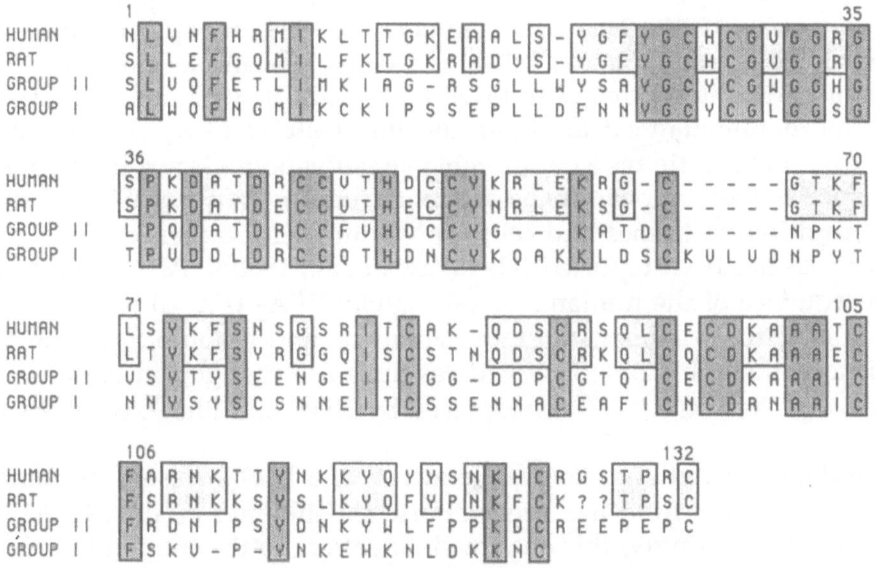

Fig. 6.   Alignment of amino acid sequences of PLA$_2$ from human
          platelet, rat platelet (Hayakawa et al., 1989), *C.atrox* venom
          (Group II) and bovine pancreas (Group I). Identical residues are
          boxed (dark shading: homology between all PLA$_2$s; light shading:
          sequence identity between human and rat platelet PLA$_2$).

positively charged residues (Arg[54], Lys[57] and Arg[58]), consistent with the observed preference for negatively charged phospholipid substrates. As reported recently, autocatalytic acylation of Lys-residues near the amino-terminus or within the "loop" results in enzyme dimerization and may play a role in the activation of monomeric PLA$_2$s (Cho et al., 1988; Tomasselli et al., 1989). Finally, five positively charged residues (Arg[54], Lys[57], Arg[58], Lys[69] and Lys[74]) alternating with hydrophobic residues are present between residues 54 and 77, identified as the "anticoagulant region" of PLA$_2$s (Kini and Evans, 1987). The presence of positively charged residues (and overall positive charge) in this segment correlates with the anticoagulant potency of PLA$_2$s and is highest (five charged residues) for crotoxin B, a PLA$_2$ with a strong anti-coagulant nature from the venom of *C.d.terrificus*. Since the human PLA$_2$ has a similar positive charge distribution in this section, it may have anticoagulant properties.

The existence of mammalian PLA$_2$s with structural similarity to Group II enzymes has been suggested from amino-terminal sequences of PLA$_2$s from porcine intestine (Verger et al., 1982), rabbit and rat inflammatory exudate (Forst et al., 1986; Chang et al., 1987a), rat and rabbit platelet (Hayakawa et al., 1987, 1988a; Mizushima et al., 1989), rat spleen (Ono et al., 1988), rat liver (Aarsman et al., 1989), as well as human placenta and rheumatoid arthritic synovial fluid (Lai and Wada, 1988; Hara et al., 1988). To date all mammalian extracellular and intracellular PLA$_2$s isolated are either of the pancreatic type or resemble snake venom Group II enzymes. Within species there appears to be sequence identity between PLA$_2$s isolated from different tissues and cells. The amino acid sequence of rat platelet PLA$_2$ has been reported (Hayakawa et al., 1988b). Alignment of the protein sequence of the human and rat platelet PLA$_2$ (Fig. 6) reveals many identical structural segments and an overall sequence homology of 69%. Furthermore, some highly conserved residues are changed in both enzymes, e.g. Tyr[28] and Tyr[75] are replaced by His[28] and Phe[75], respectively.

Recently, full length amino acid sequence of a PLA$_2$ present in human rheumatoid arthritic fluid has been reported (Seilhamer et al.,1989; Kramer et al., 1989). Interestingly, this enzyme is identical to the human platelet enzyme.

Phospholipases A$_2$ and disease

There is significant evidence to indicate that PLA$_2$s are involved in the pathogenesis of many diseases. Thus, local and circulating levels of PLA$_2$ enzyme and reaction products are elevated during infections, inflammatory diseases, tissue injury and brain dysfunction (Vadas and Pruzanski, 1986; Chang et al., 1987b; Farooqui et al., 1987) and correlate with the severity, magnitude and duration of theses disorders (Gattaz et al., 1986; Vadas and Pruzanski, 1986; Vadas et al., 1988a and 1988b; Pruzanski et al., 1988). Uncontrolled or excessive PLA$_2$ activity may promote chronic inflammation, allergic reactions, tissue injury and pathophysiological complications. These effects may be the result of accumulating PLA$_2$ products (lysophospholipids and free fatty acids) and destruction of key structural phospholipid components, but undoubtedly are potentiated by secondary metabolites, such as eicosanoids and platelet-activating factor. PLA$_2$ products or lipid mediators derived thereof have been implicated in numerous activities that are an integral part in effector cell activation, chemotaxis, adhesion, degranulation, phagocytosis and aggregation (O'Flaherty, 1982; Zimmermann et al., 1987). The (patho)physiological sequelae of these cellular events include changes in smooth muscle tone, vascular permeability and blood flow and result in influx of inflammatory cells, tissue damage, fever and pain (Larsen and Henson, 1983; Davies et al., 1984).

Activation of cytosolic PLA$_2$ via receptors may be a critical early event in the pathogenesis of allergic reactions. Release of histamine by mast cells and basophils as triggered by antigen-IgE requires activation of PLA$_2$ and generation of arachidonic acid metabolites (Marone et al., 1981; Nakamura and Ui, 1985; Yamada et al., 1987). Significantly, such receptor mediated histamine release is mimicked by exposure of mast cells to exogenous PLA$_2$s (Brain et al., 1977; Chi et al., 1982). During acute myocardial ischemia, on the other hand, cytosolic PLA$_2$ may be activated due to increased intracellular Ca^{2+} (Chien et al., 1979). Subsequent accumulation of lysophospholipids (Sobel et al., 1976) and free fatty acids (Chien et al., 1984) may promote damage to sarcolemmal membranes leading to irreversible cell injury and ultimately cell death. It has also been proposed that altered cytosolic PLA$_2$ activity or defects in its control and regulation may be a predisposing factor to affective illnesses, including depression (Hibbeln et al., 1989).

PLA$_2$s secreted excessively at local sites may be responsible for tissue damage common to rheumatic disorders, alveolar epithelial injury of lung diseases and reperfusion damage (Vadas and Pruzanski, 1986; Chang et al., 1986). Moreover, massive release of such PLA$_2$s into the circulation and airways may provide for the disturbed functions of the renal, circulatory and respiratory organ systems in acute pancreatitis, adult respiratory distress syndrom and septic shock. The source of locally emerging and systemically circulating PLA$_2$s is unknown. Various cells, including macrophages (Traynor and Authi, 1981), neutrophils (Lanni and Becker, 1983), platelets (Horigome et al., 1987, Kramer et al., 1989), chondrocytes (Chang et al., 1986), synoviocytes (Gilman et al., 1988), renal mesanglial cells (Pfeilschifter et al., 1989a) and smooth muscle cells (Pfeilschifter et al., 1989b) release PLA$_2$ upon stimulation. Moreover, interleukin 1, a polypeptide that is produced after infection, injury or antigenic challenge, was reported to induce cellular biosynthesis of PLA$_2$ (Chang et al., 1986). Recently, human secreted PLA$_2$ mRNA has been detected in human tonsils, kidney and placenta, as well as in cells derived from human rheumatoid arthritic synovial fluid (Fig. 5C), inflamed synovial tissue and peritoneal exudate (Seilhamer et al., 1989). Such mRNA was not detected in pancreas, spleen and liver. Although the presence of PLA$_2$ transcripts in isolated polymorphonuclear leucocytes, monocytes and macrophages remains to be demonstrated, it appears that extracellular PLA$_2$s may originate not only from platelets and cells of the myelomonocytic phagocyte system, but also from cells constituting inflamed or damaged tissues.

Direct evidence for the involvement of PLA$_2$ in the pathogenesis of disease has been obtained in studies using in vitro or animal models of disease, where administration of purified PLA$_2$s was found to cause disease-like symptoms (Vadas and Pruzanski, 1986). Exogenous PLA$_2$ was found to induce release of eicosanoids from leukocytes (Lam et al., 1988) and potentiate stimulation of neutrophils by the chemotactic peptide fMLF (Lackie and Lawrence, 1987). More recently, PLA$_2$ isolated from septic shock plasma was reported to promote systemic hypotension upon reinfusion (Vadas et al.,1988). Intra-articular injection of purified PLA$_2$ elicited acute inflammatory and destructive changes of joints (Vadas et al., 1989). PLA$_2$ was also found to induce nonspecific airway hyperreactivity that is a hallmark of asthma (Chand et al., 1988). Further, PLA$_2$ was reported to readily degrade lung surfactant, change the barrier properties of the alveolar epithelium and promote functional and morphological changes indicative of severe epithelial cell injury (Niewoehner et al., 1989). Similarly, exogenous PLA$_2$ increased intestinal mucosal permeability and injury (Otamiri, 1988) and produced renal tubular cell injury that was potentiated under hypoxic conditions (Nguyen et al., 1988).

Taken together these findings support the contention that both cytosolic and secreted PLA_2s are intimately involved in the pathogenesis of many diseases, including pulmonary dysfunction, ischemic and reperfusion injury, gastrointesinal disorders, connective tissue and skin diseases, disordered brain function and septic shock. Control of excessive PLA_2 secretion, activation and induction as well as inhibition of exorbitant enzymatic activity may therefore provide adjunctive therapy in the prevention or treatment of these disorders.

Conclusions

The human platelet PLA_2 that we characterized closely resembles PLA_2s from the venoms of snake and exhibits structural and functional properties of a secreted enzyme: 1) the deduced sequence encodes a signal peptide; 2) cells transfected with the PLA_2 gene secrete PLA_2 into the conditioned medium; and 3) the biochemical properties, in particular the dependence on millimolar Ca^{2+} for activity, reflect the characteristics of an extracellular enzyme. Undoubtedly, intracellular, cytosolic PLA_2s that are activated by submicromolar Ca^{2+} and prefer phosphatidylcholine substrates (Loeb and Gross, 1986; Kramer et al., 1988) are distinct enzymes and their structural properties remain to be elucidated.

It is well known that snake venom PLA_2s, although exhibiting a high degree of structural homology, differ greatly in their pharmacological properties. The pharmacological potency is most pronounced for basic PLA_2s. There is evidence to indicate that these pharmacological properties may not be solely due to hydrolytic activity, but also depend on structural features of the PLA_2 protein (Rosenberg, 1986). The human PLA_2 is most homologous to basic snake venom PLA_2s and the full spectrum of its biological activities remains to be determined. Nonetheless, it can be speculated that successful pharmacological intervention may require not only inhibition of catalytic activity, but also neutralization of cationic amphiphilic surfaces. A prerequisite for the design of both types of PLA_2 inhibitors is the knowledge of the three-dimensional structure of the human PLA_2. Consequently, the availability of sufficient quantities of recombinant PLA_2 for crystallographic analysis will play a key role in the design and discovery of potential drugs. It also seems of great importance to identify means for controlling expression of the human PLA_2 in disease.

Molecular PLA_2 probes will enable the study of regulation of PLA_2 at the transcriptional level. Such investigations may provide new insights into the involvement of secreted PLA_2s in the pathophysiology of chronic diseases and reveal new points of pharmacological intervention.

References

Aarsman, A.J., de Jong, J.G.N., Arnoldussen, E., Neys, F.W., Wassenaar, P.D. & van den Bosch, H. (1989) J.Biol.Chem. 264, 10008-10014

Achari, A., Scott, D., Barlow, P., Vidal, J.C., Otwinowski, Z., Brunie, S. & Sigler, P. (1987) Cold Spring Harbour Symp. Quant. Biol. 52, 441-452

Axelrod, J., Burch, R.M. & Jelsema, C.L. (1988) Trends Neurochem.Sci. 11,117-123

Blobel, G. (1988) Proc.Natl.Acad.Sci. 77, 1496-1500

Brain, S., Lewis, G.P. & Whittle, B.J.R. (1977) Br.J.Pharmacol., 440-441P

Chand, N., Diamantis, W., Mahoney, T.P. & Sofia, R.D. (1988) Br.J.Pharmacol. 94, 1057-1062

Chang, H.W., Kudo, I., Tomita, M. & Inoue, K. (1987a) J.Biochem. 102, 147-154

Chang, J., Gilman, S.C. & Lewis, A.J. (1986) J.Immunol. 136, 1283-1287

Chang, J., Musser, J.H. & McGregor, H. (1987b) Biochem.Pharmacol. 36, 2429-2436

Chi, E.Y., Henderson, W.R. & Klebanoff, S.J. (1982) Lab.Invest. 47, 579-585

Chien, K.R., Pfau, R.G. & Farber, J.L. (1979) Am. J. Pathol. 97, 505-522

Chien, K.R., Han, A., Sen A., Buja, L.M. & Willerson, J. T. (1984) Cir. Res. 54, 313-322

Cho, W., Tomasselli, A.G., Heinrikson, R.L. & Kezdy, F. (1988) J.Biol.Chem. 263, 11237-11241

Clark, M.A., Conway, T.M., Shorr, R.G.L. & Crooke, S. (1987)
J.Biol.Chem. 262, 4402-4406

Davies, P., Bailey, P.J. & Goldenberg, M.M. (1984) Ann.Rev.Immunol. 2,
335-357

Dawson, R.M.C., Hemington, N.L. & Irvine, R.F. (1983)
Biochem.Biophys.Res.Commun. 117, 196-201

Dennis, E.A. (1983) in The Enzymes (ed. Boyer, P.) Vol 16, pp. 307-353,
Academic Press, New York

Dijkstra, B.W., Kalk, K.H., Hol, W.G.J. & Drenth, J. (1981) J.Mol.Biol.
147,97-123

Dinarello, C. (1988) Faseb J. 2, 108-115

Dufton, M.J., Eaker, D. & Hider, R.C. (1983) Eur. J.Biochem. 137, 537-544

Elsbach, P., Weiss., J., Franson, R.C., Beckerdite-Quagliata, S., Schneider,
H. & Harris, L. (1979) J.Biol.Chem. 254, 11000-11009

Elsbach, P. (1980) Rev.Infect.Dis. 2, 106-128

Elsbach, P., Weiss, J. & Forst, S. (1986) in Lipids and Membranes (Op den
Kamp, J.A.F., Roelofsen, B. & Wirtz, K.W.A., eds.) pp. 259-286, Elsevier
Science Publishers B.V., Amsterdam

Farooqui, A.A., Taylor, W.A. & Horrocks, L.A. (1987)
Neurochem.Pathol. 7, 99-128

Forst, S., Weiss, J., Elsbach, P., Maraganore, J.M., Reardon, I. &
Heinriksen, R.L. (1986) Biochemistry 25, 8381-8385

Gattaz, W.F., Kollisch, M., Thuren, T., Virtanen, J.A. & Kinnunen, P.K.
(1987) J. Biol.Psychiatry 22, 421-426

Gilman, S.C., Chang, J., Zeigler, P.R., Uhl, J. & Mochan, E. (1988)
Arthritis Rheum. 31, 126-130

Hayakawa, M., Horigome, K., Kudo, I., Tomita, M., Nojima, S. & Inoue,
K. (1987) J.Biochem. 101, 1311-1314

Hayakawa, M., Kudo, I., Tomita, M. & Inoue, K. (1988a) J.Biochem.
103, 263-266

Hayakawa, M., Kudo, I., Tomita, M., Nojima, S. & Inoue K. (1988b)
J.Biochem. 104, 767-772

Heijne, G. J. (1985) Mol.Biol. 184, 99-105

Heinrikson, R.L., Krueger, E.T. & Keim, P.S. (1977) J.Biol.Chem. 252,
4913-4921

Heinrikson, R.L. (1982) in Proteins in Biology and Medicine (Bradshaw,
R.A., Hill, R.L., Tang, J., Chiah-chuan, L., Tien-chin, T. & Chen-lu, T.
eds.) pp. 131-152, Academic Press, New York

Hibbeln, J.R., Palmer, J.W. & Davis, J.M. (1989) Biol.Psychiatry 25, 945-
961

Horigome, K., Hayakawa, M., Inoue, K. & Nojima, S. (1987) J.Biochem.
101, 53-61

Irvine, R.F. (1982) Biochem.J. 204, 3-16

Keith, C., Feldman, D.C., Deganello, S., Glick, J., Ward, K.B., Jones, E.O.
& Sigler, P.B. (1981) J.Biol.Chem. 256, 8602-8607

Kini, R.M. & Evans, H.J. (1987) J.Biol.Chem. 262, 14402-14407

Kramer, R.M., Checani, G.C., Deykin, A., Pritzker, C.R. & Deykin, D.
(1986) Biochim.Biophys.Acta 878, 394-403

Kramer, R.M., Checani, G.C. & Deykin, D. (1987) Biochem. J. 248, 779-783

Kramer, R.M., Jakubowski, J.A. & Deykin, D. (1988) Biochim.
Biophys.Acta 959, 269-279

Kramer, R.M., Hession, C., Johansen, B., Hayes, G., McGray, P.,
Chow, E.P., Tizard, R. & Pepinsky, R.B. (1989) J.Biol.Chem. 264, 5768-
5775

Kuipers, O.P., Thunnissen, M.M.G.M., de Geus, P., Dijkstra, B.W.,
Drenth, J., Verheij, H.M. & de Haas G.H. (1989) Science 244, 82-85

Lackie, J.M. & Lawrence, A.J. (1987) Biochem.Pharmacol. 36, 1941-1945

Lam, B.K., Lee, C.Y. & Wong, P. Y-K (1988) Ann.N.Y. Acad.Sci. 524, 27-34

Lanni, C. & Becker, E.L. (1983) Am.J.Pathol. 113, 90-94

Larsen, G.L. & Henson, P.M. (1983) Ann.Rev.Immunol. 1, 335-359

Loeb, L. & Gross, R. (1986) J.Biol.Chem. 261, 10467-10470

Marone, G., Kagey-Sobotka, A. & Lichtenstein, L.M. (1981)
Clin.Immunol.Immunopathol. 20, 231-239

Mizushima, H., Kudo, I., Horigome, K., Murakami, M., Hayakawa, M.,
Kim., D., Kondo, E., Tomita, M. & Inoue, K. (1989) J.Biochem. 105, 520-
525

Nakamura, T. & Ui., M. (1985) J.Biol.Chem. 260, 3584-3593

Nguyen, V.D., Cieslinski, D.A. & Humes H.D. (1988) J.Clin.Invest. 82,
1098-1105

Niewoehner, D.E., Rice K., Duane, P., Sinha, A.A., Gebhard, R. &
Wangensteen, D. (1989) J.Appl.Physiol. 66, 261-267

O'Flaherty, J.T. (1982) Lab.Invest. 47, 314-329

Ono, T., Tojo, H., Kuramitsu, S., Kagamiyama, H. & Okamoto, M. (1988)
J.Biol.Chem. 263, 5732-5738

Otamiri, T. (1988) Agents and Actions 25, 378-384

Pfeilschifter, J., Pignat, W., Vosbeck, K. & Marki, F. (1989a)
Biochem.Biophys.Res.Commun. 159, 385-394

Pfeilschifter, J., Pignat, W., Marki, F. & Wiesenberg, I. (1989b)
Eur.J.Biochem. 181, 237-242

Pruzanski, W., Keystone, E.C., Sternby, B., Bombardier, C., Snow, K.M.
& Vadas, P. (1988) J.Rheumatol. 15, 1351-1355

Renetseder, R., Brunie, S., Dijkstra, B.W., Drenth, J. & Sigler, P.B. (1985) J.Biol.Chem. 260, 11627-11634

Rosenberg, P. (1986) in Natural Toxins (Harris, J.R., ed.) pp. 129-174, Oxford University Press, Oxford

Seilhamer, J.J., Randall, T.L., Yamanaka, M. & Johnson, L.K. (1986) DNA 5, 519-527

Seilhamer, J.J., Pruzanski, W., Vadas, P., Plant, S., Miller, J.A., Kloss, J. & Johnson, L.K. (1989) J.Biol.Chem. 264, 5335-5338

Slotboom, A.J., Verheij, H.M. & de Haas, G.H. (1982) in New Comprehensive Biochemistry (Neuberger, A. & van Deenen, L.L.M., eds) Vol 4, pp. 359-434, Elsevier Science Publishers B.V., Amsterdam

Snyder, F. (1985) Med.Res.Rev. 5, 107-140

Sobel, B.E., Corr, P.B., Robinson, A.K., Goldstein, R.A., Witowski, F.X. & Klein, M. S. (1978) J. Clin. Invest. 62, 546-533

Tomasselli, A. G., Hui, J., Fisher, J., Zurcher-Neely, H., Reardon, I.M, Oriaku, E., Kezdy, F.J. & Heinrikson, R.L. (1989) J.Biol.Chem. 264, 10041-1004

Traynor, J.R. & Authi, K.S. (1981) Biochim.Biophys.Acta 665, 571-577

Vadas, P. & Pruzanski, W. (1986) Lab.Invest. 4, 391-404

Vadas, P., Pruzanski, W., Stefanski, E., Sternby, B., Mustard, R., Bohnen, J., Frazer, I., Farewell, V. & Bombardier, C. (1988a) Crit.Care Med. 16, 1-7

Vadas, P., Pruzanski, W. & Stefanski, E. (1988b) Agents and Actions 24, 320-325

Vadas, P., Pruzanski, W., Kim, J. & Fornasier, V. (1989) Am.J.Pathol. 134, 807-811

Van den Bosch, H. (1980) Biochim.Biophys.Acta 604, 191-246

Verger, R., Ferrato, F., Mansbach, C.M. & Pieroni, G. (1982) Biochemistry 21, 6883-6889

Verheij, H.M., Slotboom, A.J. & de Haas, G.H. (1981) Rev.Physiol. Biochem.Pharmacol. 91, 91-203

Waite, M. (1987) in Phospholipases (Hanahan, D.J., ed) Plenum Publishing Corp., New York

Yamada, K., Okano, Y., Miura, K. & Nozawa, Y. (1987) Biochem.J. 247, 95-99

Zimmermann, G.A., Whatley, R.E., McIntyre, T.M. & Prescott, S.M. (1987) Am.Rev.Respir.Dis. 136, 204-207

PURIFICATION AND CHARACTERIZATION OF A PHOSPHOLIPASE A$_2$ FROM HUMAN OSTEOARTHRITIC SYNOVIAL FLUID

Thomas P. Parks, Susan Lukas, and Ann F. Hoffman

Department of Biochemistry
Boehringer Ingelheim Pharmaceuticals, Inc.
90 East Ridge, Ridgefield CT 06877

SUMMARY

Phospholipase A$_2$ (PLA$_2$) from human osteoarthritic synovial fluid was purified to homogeneity in three steps. The NH$_2$-terminal amino acid sequence and biochemical characteristics of the enzyme were identical to the Peak A PLA$_2$ activity of rheumatoid synovial fluid (1). The enzyme exhibited an apparent mass of 14,000, an absolute Ca^{++}-dependence, an alkaline pH optimum, and was inhibited by sodium deoxycholate (DOC), NaCl and 0.5 M Tris-HCl. The enzyme strongly prefered phosphatidylethanolamine (PE) as substrate over phosphatidylcholine (PC) or phosphatidylinositol (PI), and hydrolyzed PE containing arachidonic acid or linoleic acid in the sn-2 position at similar rates. Heparin bound to the enzyme but did not inhibit catalytic activity. In addition, the human enzyme was not inhibited by the acidic 'chaperone' subunit of crotoxin despite considerable sequence similarity with the basic PLA$_2$ subunit of the neurotoxin. The enzyme was capable of hydrolyzing E. coli membrane phospholipids in the presence of the neutrophil bactericidal/permeability increasing protein (BPI). This finding, coupled to the reported pro-inflammatory activity and presence of the enzyme in inflammatory cells, supports the hypothesis that it may be a component of the host defense mechanism which can, under certain conditions, contribute to the pathogenesis of inflammatory disease.

INTRODUCTION

Phospholipases A_2 are a group of generally small, Ca^{++}-depen-
dent enzymes which catalyze the hydrolysis of the sn-2 acyl ester
bond of phosphoglycerides to produce lysophospholipids and free
fatty acids (2). The enzymes are widespread in nature and occur in
both cell-associated and extracellular forms. The cellular enzymes
serve important physiological roles in phospholipid metabolism,
and in membrane remodeling and repair (3-5). PLA_2 also plays a key
role in regulating the biosynthesis of inflammatory lipid mediators, including eicosanoids and platelet activating factor (PAF),
by liberating their rate-limiting precursors, arachidonic acid and
lyso-PAF, respectively (6-8).

Extracellular PLA_2s are particularly abundant in mammalian
pancreatic tissue and in the venom of snakes and arthropods, where
they serve a digestive function. The venom enzymes additionally
exhibit a wide range of toxic effects which may or may not depend
on their enzymatic activity (9,10). Increasing evidence indicates
that mammalian extracellular PLA_2s may contribute to the pathogenesis of inflammatory disease (11). Circulating PLA_2 levels are
greatly elevated during endotoxin shock in rabbits (12) and septic
shock in man (13) and correlate directly with the magnitude and
duration of hypotension. The administration of purified extracellular PLA_2 produces hypotension in experimental shock models
(11,12). PLA_2 has been found in the cell-free inflammatory peritoneal exudates of rabbits (14) and rats (15) and in the synovial
fluid of patients with inflammatory arthritides (16). The circulating levels of PLA_2 are also elevated in patients with rheumatoid arthritis, reflecting the systemic nature of the disease, and
correlate significantly with the PLA_2 activity in synovial fluid
and with clinical and laboratory indices of disease activity (17).
Intraplantar injection of partially purified human synovial fluid
PLA_2 induces edema in mice (18), and intra-articular injection of
the enzyme in rats causes inflammatory and proliferative changes
resembling those seen in experimental models of inflammatory
arthritis and in rheumatoid arthritis (19).

Despite the growing interest in mammalian non-pancreatic
extracellular PLA_2s, relatively little is known about the structural and functional properties of these enzymes due to their
scarcity. Recently, the extracellular PLA_2s present in rabbit (14)
and rat (15) inflammatory exudates, and secreted from rat (20) and
rabbit (21) platelets have been isolated and partially sequenced.
They were found to share considerable sequence homology with the
cellular PLA_2s purified from pig ileum (22) and rat spleen membranes (23). These enzymes all lacked a half cystine at residue
11, making them more closely related structurally to the Group II
crotalid/viperid venom PLA_2s than to the Group I enzymes of
mammalian pancreas and venoms of elapid and hydrophid snakes (24).

A PLA_2 present in human synovial fluid was recently isolated and partially sequenced (25,26), revealing it to be structurally related to other non-pancreatic extracellular PLA_2s. An enzyme with an identical NH_2-terminal sequence was purified from human placental membranes (26), again raising the question about the relationship between extracellular and cell-associated PLA_2s. Seilhamer et al.(1) recently reported that synovial fluid contains at least two distinct PLA_2 activities, denoted Peaks A and B, which can be resolved by reversed phase HPLC and distinguished by their biochemical properties.

Little is known about the cellular origins of inflammatory extracellular PLA_2s, or their structural and functional relationships to intracellular PLA_2s. In order to answer these questions, it is first necessary to obtain pure enzymes for structural studies and production of immunochemical and nucleic acid probes. In the present study, we purified to apparent homogeneity the major PLA_2 activity present in the synovial fluid of patients with osteoarthritis, and characterized some of its biochemical and functional properties.

MATERIALS AND METHODS

Materials

[1-^{14}C]Oleic acid (57 Ci/mol), 1-palmitoyl-2-[1-^{14}C]arachidonyl-phosphatidylethanolamine (52 Ci/mol), 1-palmitoyl-2-[1-^{14}C]arachidonyl-phosphatidylcholine (52 Ci/mol), 1-stearoyl-2-[1-^{14}C]arachidonyl-phosphatidylinositol (45.4 Ci/mol) and [5,6,8,-9,11,12,14,15-^{3}H(N)]arachidonic acid (94.5 Ci/mmol) were purchased from Dupont-New England Nuclear (Boston, MA). 1-Stearoyl-2-[1-^{14}C]linoleoyl-phosphatidylethanolamine (55.6 Ci/mol) was purchased from Amersham (Arlington Heights, IL). Pre-swollen, microgranular carboxymethyl-cellulose (CM-52) was from Whatman (Hillsboro, OR). Heparin-Sepharose CL-6B was obtained from Pharmacia LKB Biotechnology (Piscataway, NJ). Heparin, carrageenan and buffer salts (Bis-Tris, CAPS, CHES, Tris, Glycine) were purchased from Sigma (St. Louis, MO). Chondroitin sulfate A and B', and hyaluronic acid were from Calbiochem (San Diego, CA). Fatty acid free bovine serum albumin (FAF-BSA) was obtained from Boehringer Mannheim Biochemicals (Indianapolis, IN) and sodium deoxycholate (DOC) from Aldrich (Milwaukee, WI).

E. Coli-based PLA_2 Assay

Autoclaved,[^{14}C]oleic acid-labeled E. coli (strain W) for use as a PLA_2 substrate was prepared and characterized as described by Patriarca et al. (27). The assay mixture contained 100 mM Tris-HCl, pH 7.5, 10 mM $CaCl_2$, 1 mg/ml FAF-BSA, [^{14}C]oleic

acid-labeled *E. coli* (6.5 nmol phospholipid, 30,000 dpm), and
enzyme in a final volume of 0.5 ml. The reaction was initiated by
addition of substrate, run at room temperature for various times,
and terminated by addition of 0.1 ml of 2.5 N HCl and 0.3 ml of
33.3 mg/ml FAF-BSA. The *E. coli* were sedimented by centrifugation
at 16,000 x g for 3 min, and 0.45 ml of the supernatant removed
for counting in a Beckman model LS7500 liquid scintillation
counter. Activity was usually expressed as percent conversion,
i.e. the percentage of [^{14}C]oleic acid released from the *E. coli*
membranes.

Micelle PLA$_2$ Assay

The assay mixture contained 100 mM Tris-HCl, pH 9.0, 10 mM
CaCl$_2$, 0.5 mg/ml FAF-BSA, [^{14}C]arachidonyl-PE (1.5 nmol, 175,000
dpm) and enzyme in a final volume of 100 μl. The assay was begun
by addition of substrate (dispersed by sonication in 10 mM Tris-
HCl, pH 9.0), run for various times at 37°C, and terminated by
addition of 1.0 ml of Dole's reagent (2-propanol/heptane/0.5 M
sulfuric acid, 40:10:1, v/v/v) and fatty acids extracted as
described by Kramer et al. (28). An aliquot of the organic phase
was mixed vigorously with 150 mg of heat activated silicic acid
(Bio-Sil A, Bio-Rad, Richmond, CA), the silicic acid sedimented by
centrifugation, and the organic extract decanted into vials and
counted. To correct for recovery, 30,000 dpm of [^{3}H]arachidonic
acid was included in the assay mix and a dual label dpm program
used to quantify the [^{14}C]arachidonic acid released. Activity was
expressed as pmol/min.

Other Analytical Procedures

Protein concentration was determined using the Pierce BCA
protein assay reagent (Pierce, Rockford, IL) with bovine serum
albumin as a standard. For very dilute or small samples, protein
was determined by amino acid analysis using a Beckman System 7300
amino acid analyzer following hydrolysis under vacuum in 6 N HCl
and 2% phenol for 1 hour at 150°C. Sodium dodecylsulfate-poly-
acrylamide gel electrophoresis (SDS-PAGE) was carried out as
described by Laemmli (29) using precast 4-20% gradient gels
obtained from Novex (Encinitas, CA). Proteins were visualized
using Coomassie Blue R-250 or a silver staining kit (Bio-Rad)
Amino acid sequence analysis was performed using an automated gas
phase sequenator (Model 477A, Applied Biosystems, Foster City, CA)
and the resulting phenylthiohydantoin amino acids analyzed on-line
by HPLC with an Applied Biosystems 120A PTH analyzer.

Purification of Synovial Fluid PLA$_2$

Synovial fluids obtained from patients with osteoarthritis were centrifuged at 4°C to remove cells and debris and stored at -70°C until used. The initial dialysis/precipitation procedure was adapted from Stefanski et al. (30). Synovial fluid (195 ml) was dialyzed against 40 liters of 5 mM ammonium acetate buffer, pH 5.5, for 3 days at 4°C with 3 changes of buffer. The precipitated material was collected by centrifugation at 13,000 x g for 30 min. The pellet was dissolved in 50 ml of buffer A (50 mM ammonium acetate buffer, 0.4 M NaCl, pH 5) by stirring overnight at 4°C. The solution was clarified by centrifugation at 13,000 x g for 40 min, and the pelleted material re-extracted with 50 ml of buffer A. Following centrifugation, the supernatant fractions were pooled and loaded onto a 2.5 x 25 cm CM-52 column equilibrated in buffer A. The column was washed with buffer A at a flow rate of 25 ml/hr until the A$_{280}$ of the effluent reached baseline. PLA$_2$ activity was eluted using a 600 ml linear gradient to 100% buffer B (0.2 M ammonium acetate buffer, 1.6 M NaCl. pH 8) as described by Lai and Wada (26). The active fractions were pooled, trifluoroacetic acid (TFA) added to a final concentration of 0.1%, and loaded directly onto a 4.6 x 250 mm Vydac C$_{18}$ reversed phase column equilibrated with Buffer A (0.1% TFA, 5% acetonitrile). The column was washed with 5 ml of Buffer A, stepped up to 20% buffer B (0.1% TFA, 95% acetonitrile) over 30 ml, and a 20-40% linear gradient of Buffer B run over 90 ml at a flow rate of 1.5 ml/min. For use, aliquots of enzyme from the reversed phase (RP) column were dried *in vacuo*, and reconstituted in 10 mM Tris-HCl, pH 8.0, containing FAF-BSA. PLA$_2$ from rheumatoid synovial fluid was purified as described except a Vydac C$_4$ column was used for the HPLC step.

Purification of Crotoxin Subunits

The A and B components of crotoxin were isolated from *Crotalus durissus terrificus* venom (Sigma) as described by Aird et al. (30,31), except that the reversed phase HPLC separations were conducted on a Vydac C$_4$ column using the elution conditions described above. The NH$_2$-terminus of the B component was sequenced and found to be identical to the published sequence of Aird et al. (31) except that His-1 was replaced with Ser and Ile-18 with Val.

RESULTS

Purification

The major PLA$_2$ activity of human osteoarthritic synovial fluid was purified to apparent homogeneity using a combination of dialysis/precipitation, cation exchange chromatography, and reversed phase HPLC as summarized in Table I. About 14 μg of pure

Table I.
Purification of PLA$_2$ from human osteoarthritis synovial fluid.

Step	Protein (mg)	Specific Activity" (nmol/min/mg)
Synovial Fluid (195ml)	7,995	0.2 (0.7[+])
Dialysis/Precipitation	1,400	8.0 (1.03[+])
CM-Cellulose Column	0.092[*]	4,590
Reversed-Phase HPLC	0.014[*]	17,000

[*] Determined by amino acid analysis
[**] Determined using PE micelle assay
[+] Determined using *E. coli*-based assay

enzyme with a specific activity of 17 μmol/min/mg was obtained
from 195 ml of pooled synovial fluid. Difficulties were
encountered obtaining accurate activity measurements in the cruder
preparations, so estimates of the fold purification ranged from
24,000 to 85,000, and recoveries from 4-15%, depending on whether
the *E. coli*-based assay or the PE micelle assay was used to
determine the specific activity of PLA$_2$ in the starting material.
The initial purification step reduced the viscosity of the preparation and provided a modest increase in specific activity. A
marked increase in total activity was observed using the micelle
assay which was not evident with the *E. coli* assay. Both the
reduction in viscosity and the increase in total activity were
probably due to the removal of hyaluronic acid, which inhibited
the micelle assay to a much greater extent than the *E. coli* assay
(see below). The enzyme was retained on a CM-52 column and could
be eluted without a detectable A$_{280}$ peak, resulting in a substantial purification. Due to the tendency of the synovial fluid PLA$_2$
to adsorb to surfaces, concentration and dialysis steps and size
exclusion chromatography were avoided. The final purification was
achieved by reversed phase HPLC using a C$_{18}$ column. A sharp activity peak eluted at approximately 24% acetonitrile, corresponding to a single A$_{214}$ peak. The HPLC peak contained a single protein
band with a mass of about 14,000 as determined by SDS-PAGE and
silver staining (Figure 1). Automated Edman degradation of the
protein yielded a single, unambiguous NH$_2$-terminal amino acid
sequence of 25 residues (Figure 2). No signal was obtained on the
26th cycle, probably due to the presence of a conserved Cys

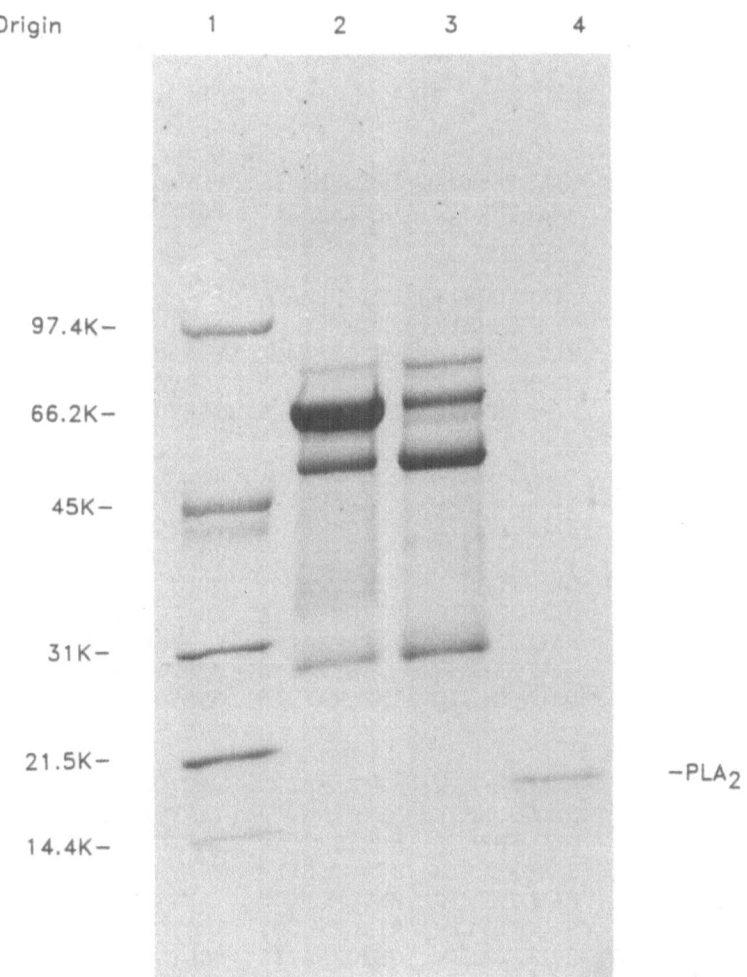

Figure 1. SDS-PAGE of osteoarthritis synovial fluid PLA$_2$ proteins. Lane 1, molecular weight marker proteins (M$_r$ indicated to the left); Lane 2, crude synovial fluid (15 μg protein); Lane 3, post-dialysis/precipitation (15 μg protein); Lane 4, post-reversed phase HPLC (1 μg PLA$_2$). The pooled CM-cellulose activity peak was not included because it was too dilute and contained too much salt. In a separate series of experiments, the synovial fluid PLA$_2$ co-migrated with the 14K pig pancreatic enzyme, and both enzymes migrated slightly slower than lysozyme (14.4K) in this electrophoresis system.

PLA₂ Source	1	10	20	Reference

**Human**

PLA₂ Source	1	10	20	Reference
<u>Human</u>				
OA Synovial Fluid	NLVNFHRMIK	LTTGKEAALS	YGFYGX	(Present study)
RA Synovial Fluid	XLVNFHRMIK	LTTGKEAALS	YGFYGX	(1)
RA Synovial Fluid	NLVNFHRMIK	LTTGKEAAL		(34)
RA Synovial Fluid*	NLVNFHRMIK	LTTG		(26)
Platelet	NLVNFHRMIK	LTTGKEAAL		(34)
Placenta	NLVNFHRMIK	LTTG		(26)
Inflam. Cell cDNA	NLVNFHRMIK	LTTGKEAALS	YGFYGC	(51)
Genomic Cloning*	NLVNFHRMIK	LTTGKEAALS	YGFYGC	(34)
Spleen	NLVNFHRMIK	LTTGKEAALS	YGFYGC	(52)
<u>Other Mammals</u>				
Rat Platelet	SLLEFGQMIL	FKTGKRADVS	YGFYGC	(20)
Rat Ascitic Fluid	XLLEFGQMIL	FKTGKRADVS	YGFYGC	(15)
Rat Spleen	XLLEFGQMIL	FKTGKRADVS	YGFYGC	(23)
Rat Liver	DLLEFGQMIL	FKTGKRADVS	YGFY	(55)
Rabbit Platelet	HLLDFRKMIR	YTTGKEATTS	YGAYGC	(21)
Rabbit Ascitic Fluid	HLLDFRKMIR	YTTGKEATTS	YGAYGC	(14)
Rabbit Leukocyte	ALLDFRKMIR	YTTGKEATXS	YGAYG	(56)
Pig Ileum	DLLNFRKMIK	LKTGKAPVPM	YAFYGC	(22)
<u>Snake Venoms</u>				
C. d. terrificus	HLLQFNKMIK	FETRKNAIPF	YAFYGC	(32)
	SLLQFNKMIK	FETRKNAVPF	YAFYGX	(Present study)
T. flavoviridis	HLLQFRKMIK	KMMTGKEPIV	SAFYGC	(58)
C. atrox	SLVQFETLIM	KIAGRSGLLW	YSAYGC	(cf. 2)
A.h.blomhoffii(acidic)	HLLQFRKMIK	KMTGKEPVIS	YAFYGC	(59)
(basic)	SLQQFETLIM	KIAGRSGIWY	SGSYGC	(39)
A.p.piscivorus (K-49)	SVLELGKMIL	QETGKNAITS	YGSYGC	(57)
(D-49)	NLFQFEKLIK	KMTGKSGMLW	YSAYGC	(57)

    source of synovial fluid not indicated
    also transfected COS-7 cell cDNA

Figure 2. The NH₂-terminal amino acid sequence of osteoarthritis
(OA), rheumatoid arthritis (RA) synovial fluid PLA₂s and selected
Group II PLA₂s.

residue in the non-reduced sample. The NH_2-terminal sequence was identical to the rheumatoid synovial fluid PLA_2 sequence first published by Hara et al. (25). As noted by Lai and Wada (26), synovial fluid PLA_2 exhibits considerable sequence similarity with other mammalian non-pancreatic PLA_2s and basic crotalid snake venom PLA_2s, particularly the B component of crotoxin. Several Group II mammalian and snake venom PLA_2 sequences are shown in Figure 2 for comparison.

Characterization

Calcium Dependence

Synovial fluid PLA_2 exhibited an absolute dependence on Ca^{++} for activity (Figure 3). In the micelle assay, the enzyme was inactive in the absence of exogenous Ca^{++} and maximally active at 5 mM Ca^{++} (Figure 3A). Activity decreased at higher Ca^{++} concentrations. In the *E. coli* assay, the enzyme exhibited considerable activity in the absence of exogenous Ca^{++} (>50% of maximal activity), probably due to Ca^{++} associated with the *E. coli* membranes (32), and was insensitive to higher Ca^{++} concentrations (Figure 3B). The residual activity observed in Ca^{++}-free buffer was abolished with EGTA. Other divalent cations (Mg^{++}, Mn^{++}, Sr^{++}, Ba^{++}, Cu^{++}, and Zn^{++}) and La^{3+} were unable to restore enzymic activity in the absence of Ca^{++} (Table II). In addition, Mn^{++}, La^{3+},

Table II. Effect of multivalent cations on synovial fluid PLA_2 activity.

Addition	$-CaCl_2$	$+5mM$ $CaCl_2$
None	0	100
$BaCl_2$	<1	61
$CuCl_2$	<1	80
$LaCl_3$	<1	10
$MgCl_2$	<1	98
$MnCl_2$	<1	7
$SrCl_2$	<1	63
$ZnCl_2$	<1	10

Data are expressed as the percent of the activity in the presence of 5 mM $CaCl_2$, using the PE micelle assay.

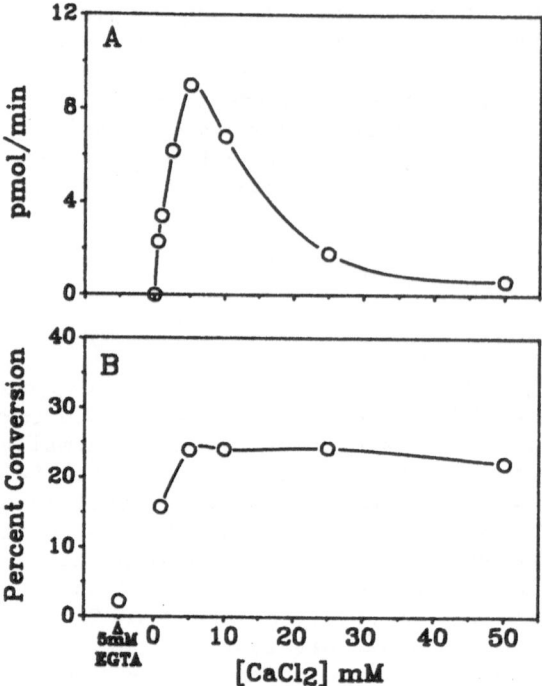

Figure 3. Calcium dependence of synovial fluid PLA$_2$. (A) Micelle
assay, (B) E.coli-based assay

and Zn^{++} suppressed activity in the presence of Ca^{++} (Table II).

pH Dependence

Synovial fluid PLA$_2$ was most active in the neutral to
alkaline pH range, depending on which assay system was employed
(Figure 4). The pH optimum was 9.5 and rather narrow using PE
micelles as substrate (Figure 4A), whereas it was approximately
7.0-7.5 and broad using autoclaved E. coli (Figure 4B). The
micelle assay was affected to a greater extent than the E. coli
assay by the different buffers used to construct the pH profile,
possibly due to differences in the ionic strength of the buffers
or to the buffer salts themselves.

Substrate Specificity/Surfactant Effects

Earlier work indicated that the PLA_2 activity in synovial fluid hydrolyzed PC micelles or PC/DOC mixed micelles very poorly. In contrast, autoclaved *E. coli* membranes (primarily radiolabeled PE) were readily hydrolyzed by the enzyme, and thus provided the basis for a sensitive assay system. Using pure phospholipids, the enzyme exhibited a marked preference for PE micelles as substrate,

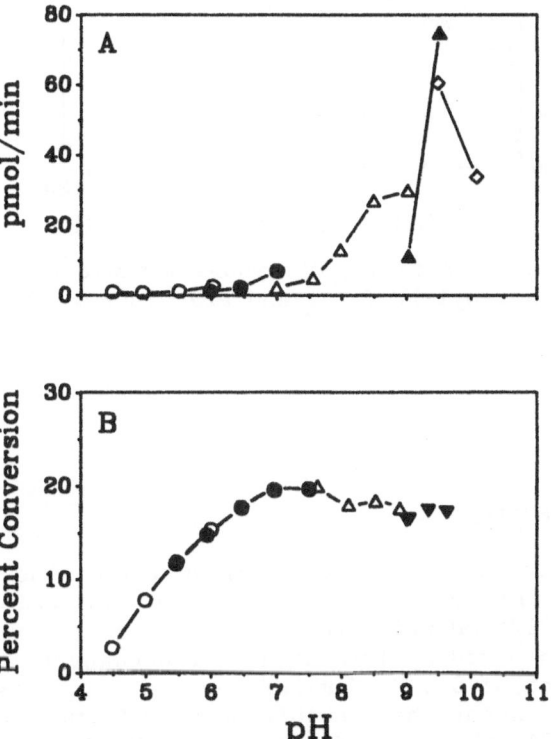

Figure 4. pH Dependence of synovial fluid PLA_2. (A) Micelle assay, (B) <u>E. coli</u>-based assay. Buffers used (0.1M): ○ Acetate, ● Bis-Tris, △ Tris, ▼ Glycine, ▲CHES, ◇ CAPS.

with the hydrolysis of PC micelles barely detectable in a 5 min
assay (Figure 5). In addition, PE containing arachidonic acid or
linoleic acid in the *sn*-2 position was hydrolyzed at a similar
rate over a wide range of substrate concentrations (Figure 5). A
low level of PC and PI hydrolysis could be observed using longer
reaction times (e.g. 60 min). The presence of 0.05% DOC did not
affect the hydrolysis of PC, but potently inhibited hydrolysis of
PE (Figure 6) and PI (not shown). Sodium dodecylsulfate and
Triton-X-100 completely inhibited PLA$_2$ activity at 0.05%.

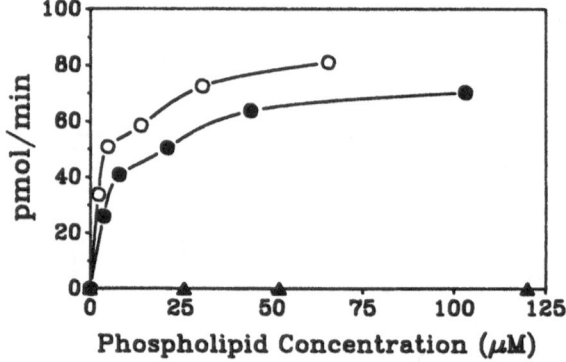

Figure 5. Substrate specificity of synovial fluid PLA$_2$. Arachidonyl-PE (O), linoleoyl-PE (●), and arachidonyl-PC (▲) were
used as substrates.

Monovalent Cation Effects

NaCl inhibited the PLA$_2$ activity of purified synovial fluid
PLA$_2$ in a concentration-dependent manner, with 50% inhibition
occurring at about 125 mM using the micelle assay (Figure 7). The
enzyme was also inhibited by Tris at higher concentrations, i.e.
about 50% inhibition at 0.5 M. Similar results were obtained using
the *E. coli* assay, except that inhibition by NaCl was less
pronounced, i.e. 50% inhibition at about 200 mM (data not shown).
Stimulation of PLA$_2$ activity by NaCl or Tris was not observed
regardless of the assay system (micelle or *E. coli*), the source of
enzyme (osteoarthritic or rheumatoid synovial fluid), or the
purity or type of the enzyme preparation used (dialysis/precipita-
tion versus acid extraction, see reference 35).

Effects of Mucopolysaccharides

Synovial fluid PLA$_2$ bound to, and could be eluted from immobilized heparin with 1.5 M KCl, as shown in Figure 8. Subsequent experiments indicated that crude or purified PLA$_2$ adsorbed to a TSK Heparin-5PW column at pH 5.0 and eluted at 0.96 M NaCl using a 0-1.6 M NaCl gradient. Soluble heparin inhibited the enzyme in a concentration-dependent manner using the *E. coli*-based assay (Figure 9B). All heparins tested had little effect, or even stimulated activity at higher concentrations in the micelle assay (Figure 9A). These results indicated that heparin binding per se

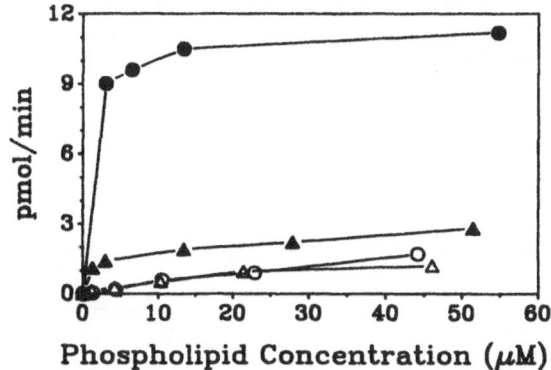

Figure 6. Effect of DOC on arachidonyl-PE (closed symbols) and arachidonyl-PC (open symbols) hydrolysis. Control reactions (●, ○); 0.05% DOC (▲, △).

did not block catalytic activity, but may have restricted access of the enzyme to the phospholipids of *E. coli* membranes. Chondroitin sulfate A was approximately 10-fold less inhibitory than heparin, and hyaluronic acid did not affect PLA$_2$ activity in the *E. coli*-based assay up to 1 mg/ml. Hyaluronic acid and chondroitin sulfate A both inhibited PLA$_2$ >50% at 100 μg/ml in the micelle assay. Another sulfated polysaccharide, carrageenan, exhibited inhibitory activity in the 1-10 μg/ml range in both assay systems. Vadas et al. (36) reported that neither heparin nor chondroitin sulfate affected the PLA$_2$ activity of crude synovial fluid using an *E. coli*-based assay.

Effect of Crotoxin 'Chaperone' Subunit

Crotoxin is a heterodimeric neurotoxin consisting of an
acidic, non-toxic subunit (component A or crotapotin) and a basic
toxic subunit possessing PLA_2 activity (component B). Crotapotin
inhibits the PLA_2 activity of component B and potentiates its
toxic effects by functioning as a 'chaperone',i.e. preventing non-
specific adsorption and thereby enhancing the specificity of
component B (37). In light of the primary sequence similarity
between the human extracellular PLA_2 and the basic PLA_2 of crotoxin, it was of interest to determine whether crotapotin was capable
of inhibiting the human enzyme. Crotapotin, which did not possess
intrinsic PLA_2 activity, inhibited the isolated crotoxin PLA_2 in a
dose-dependent manner (50% inhibition at 5 μM) but did not affect
the activity of the human enzyme using either assay system (data
not shown). This result suggests that the structural similarities
between the enzymes is not great enough for the crotoxin
'chaperone' subunit to functionally interact (i.e. inhibit
activity) with human PLA_2.

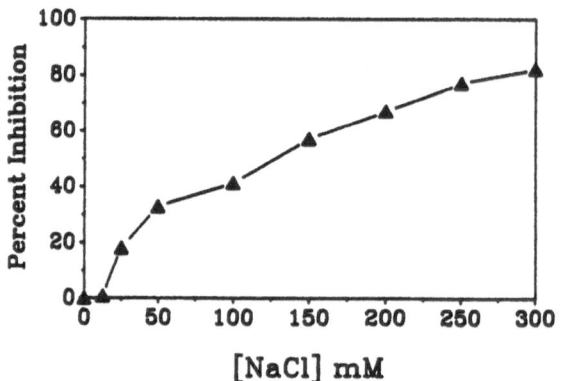

Figure 7. Inhibition of synovial fluid PLA_2 by NaCl (micelle
assay).

BPI Responsiveness

Intact *E. coli* are highly resistant to attack by PLA_2s. In
the presence of the bactericidal permeability increasing (BPI)
protein derived from neutrophils, *E. coli* membrane phospholipids

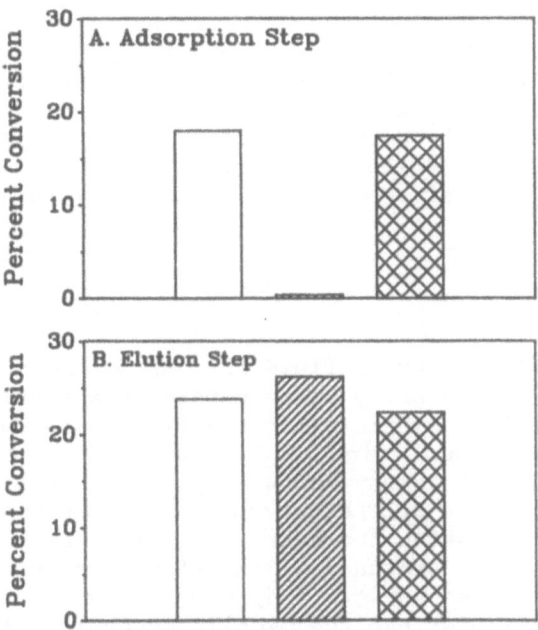

Figure 8. Synovial Fluid PLA$_2$ Binding to Heparin. (☐) Enzyme alone;(▨) enzyme + heparin-Sepharose beads;(▩) enzyme + Sepharose beads.

A. Adsorption step. Enzyme in 0.1 M Tris-HCl, pH 7.5, containing 1 mg/ml FAF-BSA was incubated alone, with heparin-Sepharose beads, or with Sepharose beads. Following centrifugation, aliquots of the supernatants were assayed for PLA$_2$ activity using the E. coli assay. The heparin-Sepharose beads had adsorbed virtually all of the PLA$_2$ in the sample.

B. Elution Step. PLA$_2$ was adsorbed to heparin-Sepharose beads, the beads washed with the above buffer, and then incubated with buffer containing 1.5 M KCl. As controls, enzyme (same amount as added to the heparin-Sepharose beads) in buffer containing 1.5 M KCl was incubated alone or with Sepharose beads. Following centrifugation, aliquots of the supernatant were assayed for PLA$_2$ activity using the E. Coli assay. In 1.5 M KCl, PLA$_2$ did not bind to heparin and all of the enzyme was recovered.

can be hydrolyzed in a highly selective fashion by certain PLA$_2$s. Comparison of the NH$_2$-terminal amino acid sequences of BPI-responsive and nonresponsive PLA$_2$s (38) and chemical modification studies (39) have suggested that this region of the PLA$_2$ molecule plays a role in BPI-dependent binding and hydrolysis. The synovial fluid PLA$_2$ was found to degrade phospholipids of *E. coli* in a BPI-dependent manner with activity comparable to that of the rabbit ascitic fluid PLA$_2$ (unpublished observations in collaboration with Dr. P. Elsbach, see reference 14). Although the synovial fluid PLA$_2$ possesses a basic NH$_2$-terminus, it lacks Arg-6 common to other known BPI-responsive enzymes, and contains an Arg at residue 7 rather than Lys.

DISCUSSION

        Osteoarthritis synovial fluid PLA$_2$ was purified in a three step procedure to homogeneity as evidenced by a single peak on reversed phase HPLC, a single band on a silver stained SDS gel, and a single NH$_2$-terminal amino acid sequence. We estimated that the enzyme used in this study had been purified at least 24,000-fold. The determination of PLA$_2$ specific activity in crude synovial fluid, upon which the degree of purification was based, was complicated by the presence of interfering substances which differentially affected the two PLA$_2$ assays. The *E. coli*-based assay was more reliable than the micelle assay for determining the specific activity of PLA$_2$ in crude synovial fluid. It should be noted, however, that the specific radioactivity of fatty acids released from *E. coli* membranes during PLA$_2$ hydrolysis was not determined and may have been significantly different from the specific radioactivity of the total *E. coli* phospholipids.

        Most studies to date have used rheumatoid synovial fluid as the starting material for isolating human non-pancreatic extracellular PLA$_2$. The rheumatoid enzyme was initially purified 4500-fold in four steps by Stefanski et al.(30). Seilhamer et al. (1) subsequently purified this enzyme preparation an additional three-fold to homogeneity by reversed phase HPLC. The HPLC step resolved two major peaks of PLA$_2$ activity which were designated Peaks A and B according to their order of elution. Peak A was the predominant activity present in rheumatoid synovial fluid, whereas Peak B was relatively more abundant in osteoarthritic and psoriatic arthritis synovial fluid (40). Lai and Wada (26) purified PLA$_2$ from 'arthritic' synovial fluid in a three step procedure based on that of Stefanski et al. (30). Although their final preparative SDS-PAGE step would not have separated the Peak A and B activities, had they been present, these investigators obtained a single NH$_2$-terminal amino acid sequence. Recently several other groups have reported purifying rheumatoid PLA$_2$ >100,000-fold to homogeneity, using reversed phase HPLC as the final step (34,41). The most

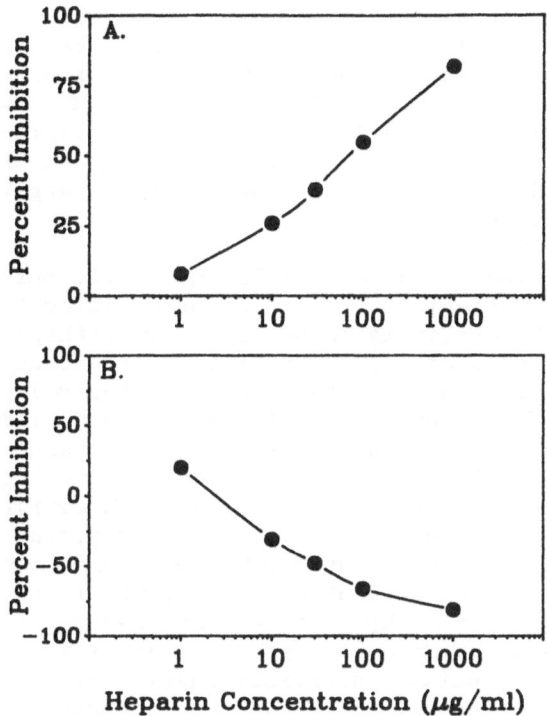

Figure 9. Effects of heparin on synovial fluid PLA$_2$ activity. (A) PE micelle assay, (B) E.coli-based assay.

striking difference between these studies and ours was the disparity in specific activities of the purified enzymes. The PLA$_2$ isolated from osteoarthritis synovial fluid exhibited a specific activity about 15-fold lower than the enzymes purified from rheumatoid synovial fluid by Kramer et al. (34) and Hara et al. (41). We have obtained specific activities >200 μmol/min/mg for PLA$_2$ isolated from rheumatoid synovial fluid, indicating that differences in assay or purification methodology were not responsible for these results. The difference in specific activities is more likely related to intrinsic differences in the underlying disease processes and compositions of the corresponding synovial fluids. Conceivably, the enzyme from osteoarthritic joints might have been partially inactivated (e.g. selective proteolysis),

and/or the rheumatoid arthritis enzyme might have been activated
(e.g. acylation, 42). In addition, we have not observed any differences in the stability of the enzyme preparations which could
explain the differences in specific activities. More work will be
required to establish whether these differences represent a
generalized feature of the PLA_2s in osteoarthritis and rheumatoid
arthritis joint fluid, and whether these results apply to other
types of arthritis.

At the time this work was completed, little information was
available on the biochemical properties of purified PLA_2 from
synovial fluid. The enzyme isolated from osteoarthritic synovial
fluid exhibited a mass of 14,000 and interacted strongly with
several different cation exchange columns at neutral pH. The
cationic nature of the enzyme greatly aided the purification
effort, but probably also contributed to its propensity to adsorb
to surfaces. This problem was circumvented by handling and storing
the enzyme in HPLC solvents, and using BSA as a carrier protein in
aqueous solutions. Stefanski et al. (30) had reported an acidic
isoelectric point for the PLA_2 partially purified from rheumatoid
synovial fluid. Based on present knowledge, their enzyme preparation either contained a highly acidic constituent, or adsorbed to
the gel and migrated anomalously during isoelectric focusing. The
Ca^{++} requirement, pH optimum, substrate specificity, and sensitivity to salt and surfactants of the osteoarthritis synovial
fluid enzyme were similar in most respects to other mammalian non-
pancreatic extracellular PLA_2s, including the enzymes purified
from rat ascitic fluid (15) and rheumatoid synovial fluid (1,30,
34,41). The osteoarthritis enzyme exhibited a much greater
specificity for PE as substrate compared to the findings of
Stefanski et al. (30), Kramer et al. (34) and Seilhamer et al. (1)
for rheumatoid PLA_2. However, this difference appeared to result
from the use of phospholipid/DOC mixed micelles as substrate by
these investigators. DOC potently inhibited the hydrolysis of PE,
but not PC, resulting in a lower PE:PC substrate preference. Our
results agreed well with those of Hara et al. (41) who used assay
methodology similar to ours.

In addition to the PE substrate specificity, the osteoarthritis enzyme did not discriminate between arachidonic acid and
linoleic acid in the sn-2 position, a property shared by PLA_2s
from rat platelets (43) and rheumatoid synovial fluid (41). In
contrast, PLA_2 activities that exhibit a selective preference for
arachidonoyl-PC have been described in human platelets (43),
neutrophils (45), macrophages (46-48), and placental blood vessels
(49). Kramer et al. (34) provided evidence that the PLA_2s from
H_2SO_4-extracted human platelets and rheumatoid synovial fluid were
identical, based on NH_2-terminal sequence data and biochemical
characteristics, and suggested that platelets may be a source of
the enzyme in arthritic joints. Although platelets are not a

common cell type in arthritic joints, and thus an unlikely source of the enzyme in synovial fluid, this study raised an intriguing question about the relationship between the PLA_2s isolated from human platelets under mild conditions versus acid-extraction. The 'native' platelet PLA_2 reportedly had an apparent molecular weight of 60,000, bound to an anion exchange column, and was activated by μM Ca^{++} concentrations (50). In addition, the PLA_2 activity present in platelet homogenates or 100,000 x g supernatants did not bind to heparin (44). The enzyme purified from acid-extracted platelets (and synovial fluid) did not exhibit any of these characteristics, suggesting that the 'native' platelet PLA_2 was either a different enzyme, or the same enzyme whose properties had been modified by oligomerization or binding to some other component.

Fawzy et al. (35) reported that cations such as Na and Tris stimulated the activity of a partially purified rheumatoid synovial fluid PLA_2 preparation using an *E. coli*-based assay. Seilhamer et al. (1,40) indicated that the Peak B PLA_2 activity was stimulated by high concentrations of Tris and by NaCl under certain conditions. In contrast we found that NaCl and Tris inhibited crude or purified PLA_2 from osteoarthritic synovial fluid using either the *E. coli* or micelle assays. Hara et al. (41) had also observed inhibition of rheumatoid PLA_2 by NaCl in a micelle assay.

The original intent of this study was to compare the structural and biochemical properties of the Peak A and B PLA_2 activities. We used osteoarthritis joint fluid as our starting material, presuming that it would contain both forms of the enzyme. We had previously encountered the Peak B activity in a sample of juvenile rheumatoid arthritis synovial fluid but not enough material was available for further investigation. In the present study, only one peak of PLA_2 activity eluted from the reversed phase column, and it became evident during the biochemical characterization of the enzyme that we had purified the peak A PLA_2, which we had also purified from rheumatoid synovial fluid. The osteoarthritis enzyme possessed an NH_2-terminal amino acid sequence identical to the Peak A PLA_2, preferred PE as substrate over PC, and was inhibited by 0.5 M Tris and 0.05% DOC. In contrast, the Peak B activity reportedly prefers PC as substrate, and was stimulated by DOC and high Tris concentrations (1,40). No sequence data for the Peak B activity has been reported yet. Why the Peak B enzyme was not detected in the present study is unclear. It may have been lost early in the purification procedure, or it may have been labile and lost activity prior to the HPLC step, which is necessary to resolve it from Peak A. A third possibility is that the levels of peak B vary greatly from patient to patient and the pooled synovial fluid which served as our starting material simply did not contain much of the Peak B activity.

Vadas et al. (36) reported that crude synovial fluid PLA$_2$ did not interact with mucopolysaccharides such as heparin or chondroitin sulfate, as assessed with an *E. coli*-based enzyme activity assay. We have shown that mucopolysaccharides can affect PLA$_2$ activity measurements in a variety of ways. Hyaluronic acid had no effect on the *E. coli* assay, yet interfered with the hydrolysis of PE micelles, presumably by interacting with the micelles. This observation suggested that caution should be exercised in determining the specific activity of PLA$_2$ in crude biological samples with a single assay system. Purified osteo-arthritis PLA$_2$ bound to immobilized heparin, and soluble heparin interfered with the *E. coli* assay but not the micelle assay. These results indicated that heparin did not bind to the active site of the enzyme or impair its catalytic mechanism. Horigome et al. (43) had reached the same conclusion in their work on the secretory PLA$_2$ from rat platelets. Heparin affinity chromatography has been used extensively to purify lipolytic enzymes, including PLA$_2$s from rat platelets (43) and more recently, rheumatoid synovial fluid (41). Another mucopolysaccharide, chondroitin sulfate A, had little effect on the *E. coli*-based assay, but inhibited hydrolysis of PE micelles. Some inhibition of *E. coli* phospholipid hydrolysis was observed with dermatan sulfate (chondroitin sulfate B), but this was probably due to the presence of contaminating heparin. Carrageenan, a sulfated polysaccharide widely used as a phlogistic agent in experimental models of inflammation, interfered with both types of assays. Thus a variety of substances which can act as cation exchangers do not interact in the same way with PLA$_2$ and/or its substrates. This conclusion has important consequences in the measurement of PLA$_2$ activity in crude biological samples, in the use of sulfated polysaccharides in experimental inflammation models, and in the biological activities of extracellular PLA$_2$ in inflammatory diseases of man and animals. Mucopolysaccharides and proteoglycans are constituents of joint fluid, connective tissue, and basement membranes of the vascular endothelium and undoubtedly influence the distribution, biological activity, and fate of PLA$_2$ in the synovial cavity and elsewhere. It is possible, for example, that PLA$_2$-induced edema may be due in part to mast cell degranu-lation, and the heparin which is released can bind to PLA$_2$, and affect its biological activities and distribution at the site of inflammation. Further studies on heparin and other glycosaminogly-cans are needed to resolve these issues.

At the present time little is known about the cellular origins of the PLA$_2$ activities found in synovial fluid, what their pathophysiological roles are, and whether they act alone, or in concert with other activator or 'chaperone' proteins. Spontaneous or stimulated release of PLA$_2$ activity has been reported from various myeloid-derived inflammatory cells and from cells of articular structures such as chondrocytes (see reference 11). Recently Seilhamer et al. (51) and Kramer et al. (34) isolated genomic and cDNA clones encoding the Peak A PLA$_2$ sequence. Northern blot analysis indicated the presence of the PLA$_2$ mRNA in inflamed human synovial tissue, peritoneal exudate cells (mostly neutrophils) from a peritonitis patient (51), tonsil, placenta, kidney, and synovial cells from one rheumatoid patient, but not another (34). Kanda et al. (52) recently determined the primary amino acid sequence of a membrane associated PLA$_2$ from human spleen which was identical to the cDNA-deduced sequence reported by Seilhamer et al. (51) and Kramer et al. (34). Thus, evidence to date indicates that inflammatory cells, tissues undergoing inflammatory reactions, or tissues with an abundant complement of myeloid-derived cells all possess the Peak A mRNA. The Peak B PLA$_2$ activity, on the other hand, was reported to be more abundant in osteoarthritis, a degenerative disease, and thus might derive from articular structures such as cartilage. Seilhamer et al.(40) recently obtained preliminary evidence that extracts of cartilage from osteoarthritis patients contained a Tris-stimulated PLA$_2$ activity which behaved chromatographically like Peak B PLA$_2$.

The results of the present study suggest that the 'chaperone' subunit of crotoxin does not functionally interact with the human non-pancreatic PLA$_2$, i.e. it does not inhibit. It was reported earlier by Breithaupt (53) that crotapotin did not inhibit PLA$_2$s from the venoms of *Crotalus adamanteus*, *Bothrops neuwiedii*, and bee, as well as pig pancreas. However, it remains to be determined whether mammalian 'chaperone' proteins exist which might influence the distribution and biological activities of extracellular PLA$_2$. The finding that the Peak A PLA$_2$ from osteoarthritis synovial fluid is a BPI-responsive enzyme, and probably co-exists with neutrophil-derived BPI at sites of inflammation suggests that the enzyme may function in host defense against Gram-negative bacteria. Evidently its actions extend beyond killing bacteria since it causes inflammatory reactions when administered to experimental animals (17,18). It is tempting to speculate that the disregulated production and release of this enzyme may be involved in the initiation and/or propagation of inflammatory disease. It is possible that the inflammatory effects of this PLA$_2$ might be potentiated by the presence of activator proteins, such as phospholipase A$_2$ activating protein (54) which, unlike BPI, acts on mammalian cells.

ACKNOWLEDGEMENTS

We thank Craig Cardella for assistance with the *E.coli*-based PLA$_2$ assays, Peter Kinkade for HPLC advice, Richard Ingraham and Tom Seng for protein sequencing, George Korza (University of Connecticut Health Science Center, Farmington, CT) for amino acid analyses, Grace Wright, Gerrold Weiss, and Peter Elsbach (New York University School of Medicine, New York, NY) for conducting the BPI studies, and Jonathan Kay, Joyce Czop (Brigham and Women's

Hospital, Boston, MA), and Henry Esber (Mason Research Institute, Worcester, MA) for help in obtaining synovial fluids. We also thank Peter Vadas (Wellesley Hospital, Toronto, ONT) and Jeffrey Seilhamer (Ideon, Redwood City, CA) for helpful discussions and Boehringer Ingelheim Pharmaceuticals for continued support of this project.

REFERENCES

1. J.J. Seilhamer, S. Plant, W. Pruzanski, J. Schilling, E. Stefanski, P. Vadas, and L.K. Johnson. Multiple forms of phospholipase A_2 in arthritic synovial fluid. J. Biochem. 106:38-42, 1989.

2. H.M. Verheij, A.J. Slotboom, and G.H. de Haas. Structure and function of phospholipase A_2. Rev. Physiol. Biochem. Pharmacol. 91:91-203, 1981.

3. H. van den Bosch. Intracellular phospholipases A. Biochim. Biophys. Acta 604:191-246, 1980.

4. E.A. Dennis. Phospholipases.In: "The enzymes" P.D. Boyer, ed. Academic Press, New York. Vol. 16, pp. 307-353,(1983).

5. F.J.G.M. van Kuijk, A. Sevanian, G.J. Handelman and E.A. Dratz. A new role for phospholipase A_2: protection of membranes from lipid peroxidation damage. Trends. Biochem. Sci. 12:31-34, 1987.

6. M.J. Broekman. Stimulated platelets release equivalent amounts of arachidonate from phosphatidylcholine, phosphatidylethanolamine and inositides. J. Lipid Res. 27:884-891, 1986.

7. F.H. Chilton, J.M. Ellis, S.C. Olson and R.C. Wykle. 1-O-Alkyl-2-arachidonoyl-sn-glycero-3-phosphocholine: A common source of platelet activating factor and arachidonate in human polymorphonuclear leukocytes. J. Biol. Chem. 259:12014-12019, 1984.

8. D.H. Albert and F. Synder. Biosynthesis of 1-alkyl-2-acetyl-sn-glycero-3-phosphocholine (platelet activating factor) from 1-alkyl-2-acyl-sn-glycero-3-phosphocholine by rat alveolar macrophages. J. Biol. Chem. 258:97-102, 1983.

9. P. Rosenberg. The relationship between enzymatic activity and pharmacological properties of phospholipases in natural poisons. In: "Animal, Plant and Microbial Toxins" J.B.

Harris, ed. Oxford University Press, Oxford. pp.129-174 (1986).

10. R.M. Kini and H.J. Evans. A model to explain the pharmacological effects of snake venom phospholipases A_2. Toxicon 27:613-635, 1989.

11. P. Vadas and W. Pruzanski. Role of secretory phospholipase A_2 in the pathobiology of disease. Lab. Invest. 55:391-404, 1986.

12. P. Vadas and J.B. Hay. Involvement of circulating phospholipase A_2 in the pathogenesis of the hemodynamic changes in endotoxin shock. Can. J. Physiol. Pharmacol. 61:561-566, 1983.

13. P. Vadas, W. Pruzanski, E. Stefanski, B. Sternby, R. Mustard, J. Bohnen, I. Fraser, V. Farewell and C. Bombardier. Pathogenesis of hypotension in septic shock: Correlation of circulating phospholipase A_2 levels with circulatory collapse. Crit. Care Med. 16:1-7, 1988.

14. S. Forst, J. Weiss, P. Elsbach, J.M. Maraganore, I. Reardon, and R.L. Heinrikson. Structural and functional properties of a phospholipase A_2 purified from an inflammatory exudate. Biochemistry 25:8381-8385, 1986.

15. H.W. Chang, I. Kudo, M. Tomita, and K. Inoue. Purification and characterization of extracellular phospholipase A_2 from peritoneal cavity of caseinate-treated rat. J. Biochem. 102:147-154, 1987.

16. W. Pruzanski, P. Vadas, E. Stefanski and M.B. Urowitz. Phospholipase A_2 activity in sera and synovial fluids in rheumatoid arthritis and osteoarthritis. Its possible role as a proinflammatory enzyme. J. Rheumatol. 12:211-216, 1985.

17. W. Pruzanski, E.C. Keystone, B. Sternby, C. Bombardier, K.M. Snow, and P. Vadas. Serum phospholipase A_2 correlates with disease activity in rheumatoid arthritis. J. Rheumatol. 15:1351-1355, 1988.

18. B.S. Vishwanath, A.A. Fawzy, and R.C. Franson. Edema-inducing activity of phospholipase A_2 purified from human synovial fluid and inhibition by aristolochic acid. Inflammation 12:549-561, 1988.

19. P. Vadas, W., Pruzanski, J. Kim and V. Fornasier. The proinflammatory effect of intra-articular injection of

soluble human and venom phospholipase A_2. Am. J. Pathol. 134:807-811, 1989.

20. M. Hayakawa, K. Horigome, I. Kudo, M. Tomita, S. Nojima, and K. Inoue. Amino acid composition and NH_2-terminal amino acid sequence of rat platelet secretory phospholipase A_2. J. Biochem. 101:1311-1314, 1987.

21. H. Mizushima, I. Kudo, K. Horigome, M. Murakami, M. Hayakawa, D.-K. Kim, E. Kondo, M. Tomita, and K. Inoue. Purification of rabbit secretory phospholipase A_2 and its characteristics. J. Biochem. 105:520-525, 1989.

22. R. Verger, F. Ferrato, C.M. Mansbach and G Pieroni. Novel intestinal phospholipase A_2: Purification and some molecular characteristics. Biochemistry 21:6883-6889, 1982.

23. T. Ono, H. Tojo, E. Kuramitsu, H. Kagamiyama, and M.Okamoto. Purification and characterization of a membrane associated phospholipase A_2 from rat spleen. J. Biol. Chem. 263:5732-5738, 1988.

24. R.L. Heinrikson, E.T. Krueger, and P.S. Keim. Amino acid sequence of phospholipase A_2-α from the venom of *Crotalus adamanteus*. J. Biol. Chem. 252:4913-4921, 1977.

25. S. Hara, I. Kudo, K. Matsuta, T. Miyamoto, and K. Inoue. Amino acid composition and NH_2-terminal amino acid sequence of human phospholipase A_2 purified from rheumatoid synovial fluid. J. Biochem. 104:326-328, 1988.

26. C.-Y. Lai and K. Wada. Phospholipase A_2 from human synovial fluid: Purification and structural homology to the placental enzyme. Biochem. Biophys. Res. Commun. 157:488-493, 1988.

27. P. Patriarca, S. Beckerdite, and P. Elsbach. Phospholipases and phospholipid turnover in *Escherichia coli* spheroblasts. Biochim. Biophys. Acta 260:593-600, 1972.

28. R.M. Kramer, G.C. Checani, A. Deykin, C.R. Pritzker and D. Deykin. Solubilization and properties of Ca^{++}-dependent human platelet phospholipase A_2. Biochim. Biophys. Acta 878:394-403, 1986.

29. U.K. Laemmli. Cleavage of structural proteins during the assembly of the head of bacteriophage T4. Nature 227:680-685, 1970.

30. E. Stefanski, W. Pruzanski, B. Sternby, and P. Vadas.

Purification of a soluble phospholipase A_2 from synovial fluid in rheumatoid arthritis. J. Biochem. 100:1297-1303, 1986.

31. S.D. Aird, I.I. Kaiser, R.V. Lewis, and W.G. Kruggel. Rattlesnake presynaptic neurotoxins: Primary structure and evolutionary origin of the acidic subunit. Biochemistry 24: 7054-7058, 1985

32. S.D. Aird, I.I. Kaiser, R.V. Lewis, and W.G. Kruggel. A complete amino acid sequence for the basic subunit of crotoxin. Arch. Biochem. Biophys. 249:296-300, 1986.

33. P. Elsbach, J. Weiss, and L. Kao. The role of intramembrane Ca^{2+} in the hydrolysis of the phospholipids of *Escherichia coli* by Ca^{2+}-dependent phospholipases. J. Biol. Chem. 260: 1618-1622, 1985.

34. R.M. Kramer, C. Hession, B. Johansen, G. Hayes, P. McGray, E.P. Chow, R. Tizard, R.B. Pepinsky. Structure and properties of a human non-pancreatic phospholipase A_2. J. Biol. Chem. 264:5768-5775, 1989.

35. A.A. Fawzy, R. Dobrow, and R.C. Franson. Modulation of phospholipase A_2 activity in human synovial fluid by cations. Inflammation 11:389-400, 1987.

36. P. Vadas, E. Stefanski, and W. Pruzanski. Characterization of extracellular PLA_2 in rheumatoid synovial fluid. Life Sci. 36:579-587, 1985.

37. C. Bon, F. Radvanyi, B. Saliou, and G. Faure. Crotoxin: A biochemical analysis of its mode of action. J. Toxicol.-Toxin Rev. 5:125-138, 1986.

38. S. Forst, J. Weiss, J.M. Maraganore, R.L. Heinrikson, and P. Elsbach. Relation between binding and the action of phospholipases A_2 on *Escherichia coli* exposed to the bactericidal/permeability increasing protein of neutrophils. Biochim. Biophys. Acta 920:221-225, 1987.

39. S. Forst, J. Weiss, P. Blackburn, B. Frangione, F. Goni, and P. Elsbach. Amino acid sequence of a basic *Agkistrodon halys blomhoffii* phospholipase A_2. Possible role of NH_2-terminal lysines in action on phospholipids of *Escherichia coli*. Biochemistry 25:4309-4314, 1986.

40. J. Seilhamer, P. Vadas. W. Pruzanski, S. Plant, E. Stefanski, and L. Johnson. Synovial fluid phospholipase A_2 in arthritis. In:"Therapeutic Approaches to Inflammatory Disease"

A.J. Lewis, N.S. Doherty, and N.R. Ackerman, eds. Elsevier, New York. pp. 129-136 (1989).

41. S. Hara, I. Kudo, H.W. Chang, K. Matsuta, T. Miyamoto, and K. Inoue. Purification and characterization of extracellular phospholipase A_2 from human synovial fluid in rheumatoid arthritis. J. Biochem. 105:395-399, 1989.

42. W. Cho, A.G. Tomaselli, R.L. Heinrikson, and F.J. Kezdy. The chemical basis for interfacial activation of monomeric phospholipases A_2. J. Biol. Chem. 263:11237-11241, 1988.

43. K. Horigome, M. Hayakawa, K. Inoue, and S. Nojima. Purification and characterization of PLA_2 released from rat platelets. J. Biochem. 101:625-631, 1987.

44. D.K. Kim, I. Kudo, and K. Inoue. Detection in human platelets of phospholipase A_2 activity which preferentially hydrolyzes an arachidonoyl residue. J. Biochem. 104:492-494, 1988.

45. F. Alonso, P.M. Henson, and C.C. Leslie. A cytosolic phospholipase in human neutrophils that hydrolyzes arachidonoyl-containing phosphatidylcholine. Biochim. Biophys. Acta 878:273-280, 1986.

46. J. Wijkander and R. Sunder. A phospholipase A_2 hydrolyzing arachidonoyl-phospholipids in mouse peritoneal macrophages. FEBS Lett. 244:51-56, 1989.

47. I. Flesch, B. Schmidt and E. Ferber. Acyl chain specificity and kinetic properties of phospholipase A_1 and A_2 of bone marrow-derived macrophages. Z. Naturforsch. 40c:356-363, 1989.

48. C.C. Leslie, D.R. Voekler, J.Y. Channon, M.M. Wall and P.Z. Zelarney. Properties and purification of an arachidonoyl-hydrolyzing phospholipase A_2 from a macrophage cell line, RAW 264.7. Biochim. Biophys. Acta 963:476-492, 1988.

49. T.M. Karnauchow and A.C. Chan. Characterization of human placental blood vessel phospholipase A_2. Demonstration of substrate selectivity for arachidonyl-phosphatidylcholine. Int. J. Biochem. 17:1317-1319, 1985.

50. R.M. Kramer, J.A. Jakubowski, and D. Deykin. Hydrolysis of 1-alkyl-2-arachidonoyl-sn-glycero-3-phosphocholine, a common precursor of platelet activating factor and eicosanoids, by human platelet phospholipase A_2. Biochim. Biophys. Acta 959:269-279, 1988.

51. J.J. Seilhamer, W. Pruzanski, P. Vadas, S. Plant, J.A. Miller, J. Kloss, and L.K. Johnson. Cloning and recombinant expression of phospholipase A_2 present in rheumatoid arthritic synovial fluid. J. Biol. Chem. 264:5335-5338, 1989.

52. A. Kanda, T. Ono, N. Yoshida, H. Tojo and M. Okamoto. The primary structure of a membrane-associated phospholipase A_2 from human spleen. Biochem. Biophys. Res. Comm. 163:42-48, 1989.

53. H. Breithaupt. Enzymatic characteristics of crotalus phospholipase A_2 and the crotoxin complex. Toxicon 14:221-223, 1976.

54. M.A. Clark, T.A. Conway, R.G.L. Shorr, and S.T. Crooke. Identification and isolation of a mammalian protein which is antigenically and functionally related to the phospholipase A_2 stimulatory peptide melittin. J. Biol. Chem. 262:4402-4406, 1987.

55. A.J. Aarsman, J.G.N. de Jonge, E. Arnoldussen, F.W. Neys, P.D. van Wassernaar, and H. Van den Bosch. Immunoaffinity purification, partial sequence, and subcellular localization of rat liver phospholipase A_2. J. Biol. Chem. 264:10008-10014, 1989.

56. C.E. Ooi, G. Wright, J. Weiss and P. Elsbach. Purification to homogeneity and properties of rabbit granulocyte PLA_2. Clin. Res. 36:465A, 1988.

57. J.M. Maraganore, and R.L. Heinrikson. The lysine-49 phospholipase A_2 from the venom of Agkistrodon piscivorus piscivorus. Relation of structure and function to other phospholipases A_2. J. Biol. Chem. 261:4797-4804, 1986.

58. S. Tanaka, N. Mohri, H. Kihara, and Ohno, M. Amino acid sequence of Trimeresurus flavoviridis phospholipase A_2. J. Biochem. 99:281-299, 1986.

59. K. Tomoo, H. Ohishi, T. Ishida, M. Inoue, K. Ikeda, Y. Aoki, and Y. Samejima. Revised amino acid sequence, crystallization, and preliminary X-ray diffraction analysis of acidic phospholipase A_2 from the venom of Agkistrodon halys blomhoffi. J. Biol. Chem. 264:3636-3638, 1989.

PHOSPHOLIPASE A2 ACTIVATION IS THE PIVOTAL STEP IN THE EFFECTOR PATHWAY OF INFLAMMATION

Peter Vadas and Waldemar Pruzanski

Immunology Diagnostic and Research Centre
Department of Medicine, Wellesley Hospital
University of Toronto
Toronto, Ontario, CANADA   M4Y 1J3

SUMMARY

Our understanding of the mechanisms of initiation and propagation of local and systemic inflammatory processes is clearly imperfect if one uses the available therapeutic modalities as a yardstick. While glucocorticoids are potent anti-inflammatory drugs, the pharmacologic target of this class of agents has not been identified with certainty, and the use of steroids is fraught with the risk of considerable and potentially dangerous side effects. On the other hand, non-steroidal anti-inflammatory drugs (NSAIDS), while more specific, are relatively weak anti-inflammatory compounds and frequently require the addition of more potent agents. Cytotoxic drugs or anti-metabolites effectively suppress acute and chronic inflammatory reactions, but also predispose to infection and initiate the development of neoplasms following long-term exposure. The inadequacy and relative non-specificity of these approaches underscore the deficiencies in our understanding of the principles that govern these responses. A better understanding of these processes will be applicable to broad categories of human disease including autoimmunity, the collagen vascular diseases, aberrations in host defense and the response to trauma and infection.

Many of the distal mediators of inflammation, leading to direct tissue damage, are well known. These include the metabolites of arachidonic acid (prostaglandins, thromboxanes, leukotrienes, lipoxins);

kinins, histamine, the toxic oxygen radicals and
platelet activating factor (PAF). However, the
proximal mediators of inflammation responsible for
initiation of many of the cascade systems have only
recently come to light. The proximal limb of the
inflammatory response includes those autacoids that
signal the presence of microbial infection. These
signals include interleukin-1 (IL-1), tumour necrosis
factor (TNF) and phospholipase A2 (PLA2). It is
noteworthy that these factors share a multitude of
biologic activities. Moreover, in assessing the
contributions of each of these factors to inflammatory
responses in light of Koch's postulates, it is evident
that each of these factors fulfills the postulates in a
compelling manner. While these signals may
individually be significant co-contributors to the
propagation of inflammation, this review will attempt
to construct a unifying framework showing an intimate
inter-relationship amongst them. The mediators of the
proximal limb (ie pre-PLA2) and those of the distal
limb (post-PLA2) will be considered individually and
the body of evidence that unites them into a single
common effector pathway will be examined.

A.   PROXIMAL (PRE-PLA2) SIGNALS

i)   Tumour Necrosis Factor

        The existence of TNF was inferred as early as 1893
(1) with the observation of infection-associated tumour
necrosis. Over subsequent years, the concept was
elaborated by demonstration of an endotoxin shed from
the walls of gram-negative bacteria (2). Rodents
challenged with bacterial endotoxin had a circulating
serum factor which, when transferred to tumour-bearing
recipients, induced the hemorrhagic necrosis of the
tumours (3). While it became apparent that TNF
mediated the endotoxin-induced regression of tumours,
it was only with the work of Cerami (4) and Beutler (5)
that TNF was identified as the cachexia-inducing factor
of chronic infection and neoplasia.

        Human TNFa is a small non-glycosylated peptide of
157 residues with a pI of 5.3 and a single disulphide
bond bridging residues 67 and 101 (6). The gene
encoding TNFa was mapped to the short arm of chromosome
6 near the gene encoding TNFb and in close proximity
to the MHC (7). While the major source of TNFa

(cachectin)   in   man   is   the   activated
monocyte/macrophage, messenger transcript for TNFa has
been identified in natural cytotoxic cells, NK cells
and bone-marrow derived mast cells (8).

Endotoxin is the single most potent signal for TNF
transcription and translation.   TNF mRNA levels may
increase by 100-fold or more in response to LPS (9)
and extracellular secretion of the mature peptide may
increase by several thousand fold (10). TNF levels are
further augmented by IFN-γ and by GM-CSF (9), although
neither is sufficient to initiate TNF secretion in the
absence of endotoxin.

Anti-inflammatory glucocorticoids   down-regulate
gene expression of TNF both pre- and post-transcription
(10).   It may be this effect of steroids which
represents one of the major loci of anti-inflammatory
activity.   However,   the   early administration   of
steroids is required in order to achieve significant
blockade. Since peak mRNA levels are attained as early
as 30 min. after endotoxin administration, delay of as
little   as   20   min. post-LPS   results   in   loss   of
inhibitory effect (11).

The extent to which TNF is the intermediary signal
for the manifestations of endotoxemia is reflected by
the virtually complete absence of toxicity of endotoxin
in C3H/HeJ mice (12). Due to a single autosomal
mutation at the lpsd allele on chromosome IV, C3H/HeJ
mice are unable to secrete TNF in response to
endotoxemia, suggesting that TNF (and not LPS) is the
injurious   agent.   However,   the   administration   of
exogenous TNF will reproduce many of the manifestations
of endotoxemia.   Intravenous infusion of TNF induces
hypotension,   generalized   increased   vascular
permeability with resultant hemoconcentration, acute
lung   injury,   acute   tubular   necrosis,
hypertriglyceridemia and intestinal necrosis (reviewed
in 13, 14). In phase I clinical trials of recombinant
human TNF in patients with metastatic cancer (15), the
commonly encountered adverse effects were fever,
chills, non-specific constitutional symptoms and in
10% of patients, acute respiratory insufficiency.

Healthy human volunteers challenged with E. coli
endotoxin 0127:B8 consistently developed constitutional
symptoms as well as circulatory instability which
qualitatively   resembled   the   hemodynamic   changes
characteristic of septic shock (16). TNF levels rose

8-fold above controls, with peak levels at 90 - 120 min. post-endotoxin and a return to baseline levels by 4 hrs. (17). Ibuprofen abrogated the fever, tachycardia and release of stress hormones after endotoxin, suggesting involvement of the cyclo-oxygenase pathway. The half-life of TNF in rabbits is 6 - 7 min. and in man is 14 - 18 min. (18, 15). Its clearance from the circulation is regulated by receptor-mediated endocytosis.

The hemodynamic and counter-regulatory responses evoked by endotoxin are abolished in animals passively immunized against TNF. Mice pretreated with polyclonal anti-TNF antibodies were protected against lethal doses of LPS (19). Similarly, passive immunization of baboons with murine monoclonal anti-TNF protected against endotoxin-induced shock and multi-system organ failure (20).

ii)   Interleukin-1

Like TNF, IL-1 administration appears to reproduce many of the manifestations of endotoxemia (21). However, the role of IL-1 in the propagation of endotoxin-induced homeostatic changes is unclear. IL-1a and IL-1b are structurally related proteins of M_r 17500. The genes encoding IL-1 are found on chromosome 2. IL-1b is the predominant form in human monocytes (22). Unlike TNF, the exon encoding mature IL-1 is not preceded by a propeptide signal sequence for extracellular export (21). Hence, most newly synthesized IL-1 remains cell-associated (23). While endotoxin is a potent signal for IL-1 synthesis (24) as it is for TNF, serum levels of IL-1 do not measurably increase in man following experimental endotoxin challenge (17), although the presence of circulating IL-1 has been correlated with a fatal outcome in meningococcemia (25).

IL-1 derives from a wide spectrum of cellular sources. These include mononuclear phagocytes, lymphocytes of both T and B lineages, vascular smooth muscle and endothelial cells and fibroblast-like cells (21). In addition to LPS, IL-1 synthesis may be induced by TNF, GM-CSF, IFN-γ and C5a (26,27). Glucocorticoids block both the transcription of IL-1 mRNA and post-transcriptional IL-1 synthesis via cAMP (28).

The parenteral administration of IL-1 in experimental animals results in hypotension (26,29), fever (30), endothelial cell-leukocyte adhesion (31) with the resultant sequestration of circulating leukocytes in the pulmonary vascular bed and acute lung injury (32). While IL-1 is clearly not as injurious as TNF, coadministration of the two cytokines results in a marked synergism of effects (26, 32). Indeed, when sub-threshold amounts of IL-1 or TNF were infused in rabbits, no hemodynamic changes were observed (32). The combination of the two agents led to profound shock. Everaerdt et al (33) have suggested that endogenous IL-1 may act as a priming or sensitizing agent to the actions of TNF. Therefore, in a clinical setting, low grade endotoxemia or bacteremia may be sufficient to cause the synthesis of a sensitizing (but sub-clinical) amount of IL-1, which if followed by a second bout of endotoxemia, will cause the release of circulating TNF in an appropriately primed individual, leading to the usual manifestations of septic shock.

B. DISTAL (POST-PLA2) SIGNALS

i) Platelet Activating Factor

Platelet activating factor or PAF-acether is a phospholipid-derived terminal mediator of inflammation. The precursor of PAF-acether, alkyllysoglycerophosphocholine (alkyllyso-GPC) is formed by deacylation of the parent compound, 1-alkyl-2-acyl-GPC, by phospholipase A2, with subsequent acetylation by acetyltransferase. In fact, the simultaneous formation of PAF-acether and arachidonate metabolites is linked through the common intermediate, 1-alkyl-2-arachidonoyl-sn-GPC (34). This arachidonyl- -containing intermediate may be an obligatory precursor of PAF-acether in platelets and stimulated polymorphonuclear leukocytes (35). An alternative pathway of PAF-acether biosynthesis involves. CDP-choline and 1-O-alkyl-2-acetyl-sn-glycerol as precursors (36).

While many cell types are capable of synthesizing and releasing PAF-acether, the principle contributors in shock are still undefined (37). It is produced by both polymorphonuclear and mononuclear phagocytes, mast cells, basophils, eosinophils, platelets, endothelial cells and even gram-negative bacteria (38). As with TNF and IL-1, endotoxin has proven to be a major

stimulus for PAF generation (39). Rats challenged with
Salmonella enteritidis endotoxin responded with a rise
in serum PAF levels of greater than 100-fold over a 10
min. period (39). Other stimuli include IgE-binding
antigens, anti-IgE, calcium ionophores, aggregated IgG,
C5a and C5a-des-Arg, fMLP, intact bacteria, opsonized
zymosan and immune complexes (reviewed by Braquet and
Rola-Pleszczynski [37]). Some of the important
biologic effects of TNF, such as hypotension and
ischemic bowel necrosis, are reversed by PAF
antagonists (40), suggesting that PAF may be a terminal
mediator of many TNF-induced events.

Infusion of PAF in canine gastric, mesenteric and
femoral arteries caused local vasodilatation with an
accompanying fall in vascular resistance (41).
Similarly, the topical application of PAF on rat
mesentery resulted in local hyperemia (42). This
effect was reversed by the PAF antagonists, BN 52021
and WEB 2086. Intravenous infusion of PAF in rats,
rabbits, dogs and guinea pigs has produced
fundamentally similar results (39). Typically, these
animals displayed profound systemic hypotension,
increased vascular permeability, hemoconcentration,
thrombocytopenia, leukopenia, myocardial depression,
bowel necrosis and metabolic derangements. These
changes are attenuated or abolished by pretreatment of
animals with PAF antagonists. Similarly, the
pathophysiologic changes of endotoxemia are also
reversed by pretreatment with PAF antagonists (39, 43,
44).

ii) Eicosanoids

A summary of the numerous studies documenting
changes in arachidonic acid metabolism during the
course of sepsis is beyond the scope of this chapter.
The extensive literature on the generation of
eicosanoids during the course of experimental
endotoxemia and clinical septic shock has been
summarized in several comprehensive reviews (45 - 49)
to which the interested reader is referred.

C. PHOSPHOLIPASE A2

PLA2 occurs in man as both a soluble, secretory
enzyme and as a membrane-associated non-secretory

enzyme. The inter-relationship between these two compartments and the extent to which a dynamic equilibrium may exist (50, 51) are currently unknown. Neither has the physiologic function of either PLA2 been satisfactorily defined. In fact, there is an ongoing debate as to which of these PLA2s may be the more relevant to pathologic processes. However, the weight of current evidence would suggest that secretory PLA2 best correlates with local and systemic inflammatory responses and reproduces the characteristic facets of these responses.

Human inflammatory (or non-pancreatic) PLA2 is a basic, calcium-requiring peptide of 124 amino acids and MW 13939 (52). It contains 7 intra-chain disulphide bonds in an extremely stable tertiary structure. A 20 residue signal sequence encoding for membrane translocation has been identified upstream from the mature coding sequence (53). Multiple forms of this enzyme are distinguishable on the basis of altered activities in the presence of detergents or Tris buffer (54). The secretory enzymes from synovial fluid, platelets and placenta are all identical (55, 56). Secretory PLA2 has been cloned and the recombinant cDNA clone has been expressed in two different systems, namely CV-1 cells infected with recombinant vaccinia virus and in Chinese hamster ovary cells (53).

Secretory non-pancreatic PLA2 derives from osteoblasts (57), chondrocytes (58), synoviocytes (59), mesangial cells (60), platelets (55), polymorphonuclear leukocytes (61) and macrophages (62). Northern blotting has identified mRNA transcript of PLA2 in inflamed synovial tissue and peritoneal cells in bacterial peritonitis (53), and in human tonsil, placenta, kidney and rheumatoid synovial fluid cells (55). Extracellular PLA2 secretion has been described for endotoxin as well as the products of gram-positive bacteria (63), viruses (64), zymosan (65), fMLP (61) and concanavalin A (62). The cytokines IL-1 and TNF both induce the synthesis and secretion of PLA2 (57 - 60); the combination of the two agents is strongly synergistic (57, 60). While glucocorticoids protect against endotoxin-induced PLA2 release in vivo (66), they have not proven to be inhibitory under defined conditions in vitro (57).

Exogenous PLA2 is vasoactive locally (67) subsequent to intradermal injection and systemically subsequent to intravenous injection in cats, dogs, rats

and rabbits (66). Concomitant with the profound and
sustained fall in mean arterial pressure is a rise in
pulmonary vascular resistance (68). Intratracheal
instillation of venom PLA2 induced acute lung injury
(69). Using an LD30 dose of Naja naja PLA2, Edelson et
al noted the development of an acute inflammatory
infiltrate with increased pulmonary vascular
permeability and formation of hyaline membranes,
leading to impaired gas exchange (70). The acute lung
injury was followed by fibrotic changes characteristic
of resolving ARDS. Injection of human synovial fluid
PLA2 in joints induced an acute synovitis (71) while
injection in mouse foot pads caused marked edema which
was attenuated by pBPB and aristolochic acid (72, 73).
The early vascular permeability has been attributed to
mast cell degranulation with release of serotonin and
formation of PAF (74).

    A large number of clinical conditions has been
surveyed (75) but the release of non-pancreatic PLA2
appears to be most marked in septic shock, with over
100-fold increases documented (63). Serum levels of
circulating PLA2 correlate with the severity and
duration of circulatory collapse in endotoxin shock in
rabbits (66) and in septic shock in man (63). During
the acute hypotensive phase of sepsis, serum PLA2
levels were highest in patients with ARDS (76) as were
PLA2 levels in bronchoalveolar lavage fluid (77).

    The mode of endogenous regulation of serum PLA2
levels is undefined, although plasma proteins
non-specifically modulate its activity (78, 79). There
is a temporal relationship between changes in serum
cortisol and PLA2 in survivors (but not non-survivors)
of sepsis (80). However, administration of large doses
of glucocorticoids during sepsis does not alter the
rate of production or clearance of PLA2 (81). Thus,
the association of PLA2 and cortisol may be
epiphenomenal rather than causal.

D.   INDUCTION OF PLA2 BY CYTOKINES

    Exposure of target cells to IL-1 and TNF results
in the formation of metabolites of arachidonic acid,
principally PGE2 (58), as well as release of
plasminogen activator and neutral proteinases (59).
Chang et al (58) described IL-1 induced activation of
cell-associated PLA2 in rabbit articular chondrocytes

as well as enhanced extracellular release. PLA2 activation was specific for IL-1; no response was seen with IL-2, IL-3 or TNF (82). Enhanced extracellular PLA2 secretion was evident by 6 hrs after stimulation. A similar response was observed with human synovial fibroblasts (59), with resultant extracellular PGE2, PLA2 and plasminogen activator secretion over 48 hrs. These results were extended by Pfeilschifter et al (60) who found that IL-1 and TNF interacted synergistically to induce PLA2 secretion from rat renal mesangial cells. Stimulated PLA2 release was blocked by actinomycin D and cycloheximide. Rat fetal calvarial osteoblasts in primary culture also constitutively secrete PLA2. Recombinant human IL-1 and TNF synergistically enhanced PLA2 secretion following a short pulse exposure (57). The cytokine effect was blocked by inhibitors of protein synthesis but not by dexamethasone. The extracellular PLA2 secreted by each of chondrocytes, fibroblasts, renal mesangial cells and osteoblasts was neutral active and calcium-dependent, as is the human non-pancreatic pro-inflammatory PLA2 (83).

Clearly, cytokines induce the synthesis and extracellular secretion of PLA2 from a variety of cell types in vitro. A number of studies suggest that PLA2 secretion is a cytokine-mediated event in vivo. New Zealand white rabbits injected with E. coli endotoxin 0127:B8 manifested a rise in serum TNF levels within 15 min of LPS injection (18). Peak serum TNF levels were seen at approximately 2 hrs after LPS injection with a return to baseline levels by 5 hrs. In an earlier study, also using E. coli endotoxin 0127:B8 in New Zealand white rabbits, increased levels of PLA2 were detected in serum by 30 - 60 min. post-LPS with ongoing increase during the 5 hr observation period (66) as shown in Figure 1. The early appearance and rapid clearance of TNF followed by increasing PLA2 levels is consistent with the in vitro observations. Identical results have been obtained in human volunteers injected with 4 ng. of E. coli endotoxin (17). TNF levels were highest at 90 min. after LPS challenge and remained elevated for 180 min. The same sera were assayed for PLA2 activity (84). PLA2 levels increased above baseline by 60 - 90 min. post-LPS, and were still increasing at the end of the 5 hr. sampling period. The changes seen in the human volunteers were identical to those seen in rabbits challenged with endotoxin.

E.   SUMMARY

Several candidates have been proposed in the
literature as plausible mediators of septic shock.
There are compelling data implicating each of these
agents as contributing to the shock syndrome.
Certainly, the entity of septic shock is a complex
syndrome in which many host defense, counter-regulatory
and other homeostatic mechanisms come into play. It
is, therefore, not unreasonable to suggest that each of
TNF, IL-1, PLA2 and PAF (inter alia) play a
contributory role.   However, one may invoke the
principle of parsimony and suggest that each of these
signals is mechanistically linked through a common
effector pathway.   Indeed, it is evident from this
review that endotoxin induces the rapid synthesis and
secretion of TNF in both animal models and in man.   TNF
in turn induces the synthesis and intravascular
secretion of PLA2 from the appropriate target cells.
The products of PLA2 hydrolysis, namely arachidonic
acid and lysoPAF serve as the immediate precursors of
the distal effector pathway of endotoxin.   Arachidonic
acid is further metabolized to generate prostaglandins
and leukotrienes while lysoPAF is acetylated by
acetyltransferase to form PAF.   This pathway is
summarized in Figure 2.   Pruzanski (83) has proposed a
functional/anatomic staging system for this effector
pathway, which allows for correlation with clinical
manifestations and outcome of specific therapeutic
interventions.

With a detailed understanding of the pathway leading
ultimately to the manifestations of shock, it should be
possible to devise effective therapeutic strategies.
Clearly, if one can anticipate the occurence of
endotoxemia or bacteremia, the neutralization of
circulating endotoxin by passive immunization with
antiserum to LPS should be efficacious, and indeed, in
selected settings this has proven to be the case (85).
Since the LPS signal is transduced by TNF, the use of
anti-TNF antiserum (or Fab' fragments) should be
equally efficacious, and is, in the experimental
setting (19, 20). However, it is usually impossible to
make a diagnosis of septicemia until the clinical
manifestations of septicemia are evident, and in these
patients, both endotoxemia and release and clearance of
TNF have already occurred.   Therefore, the next pivotal
point of intervention is directed against the
circulating PLA2, prior to generation of its bioactive

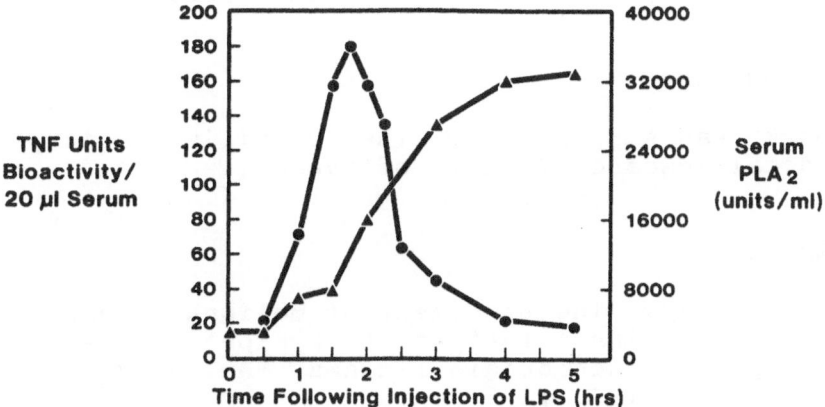

Figure 1. Effect of endotoxin challenge on serum TNF and PLA2 levels in New Zealand white rabbits.
TNF, closed circles.
PLA2, closed triangles.

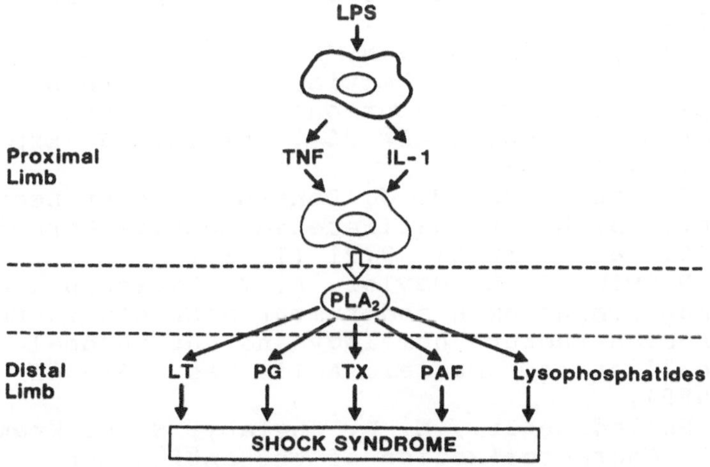

Figure 2. The proposed effector pathway of cytokines, PLA2 and terminal mediators.

products. Since secretion of PLA2 is a relatively late
event coincident with clinical deterioration, it may be
in the development of suitable pharmacologic or
biologic inhibitors of PLA2 that the dividends will be
found.

Acknowledgements

This work was supported by The Arthritis Society and
the Medical Research Council of Canada.

REFERENCES

1. W. B. Coley, The treatment of malignant tumours by
    repeated innoculations of erysipelas: With a
    report of ten original cases. Am. J. Med. Sci.
    105:487 (1893).
2. M. J. Shear, Chemical treatment of tumours. IX.
    Reactions of mice with primary subcutaneous
    tumours to injection of hemorrhage-producing
    bacterial polysaccharide. J. Natl. Canc. Inst.
    5:185 (1944).
3. E. A. Carswell, L. J. Old, R. L. Kassel, et al. An
    endotoxin-induced serum factor that causes
    necrosis of tumours. Proc. Natl. Acad. Sci.
    USA 72:3666 (1975).
4. M. Kawakami, P. H. Pekala, M. D. Lane, et al.
    Lipoprotein lipase suppression in 3T3-L1 cells
    by an endotoxin-induced mediator from exudate
    cells. Proc. Natl. Acad. Sci. USA 79:912
    (1982).
5. B. Beutler, J. Mahoney, N. Le Trang, et al.
    Purification of cachectin, a lipoprotein
    lipase-suppressing hormone secreted by
    endotoxin-induced RAW 264.7 cells. J. Exp. Med.
    161:984 (1985).
6. M. G. Rosenblum and N. J. Donato. Tumour necrosis
    factor alpha: A multifaceted peptide hormone.
    Crit. Rev. Immunol. 9:21 (1989).
7. G. E. Nedwin, S. L. Naylor, A. Y. Sakaguchi, et al.
    Human lymphotoxin and tumour necrosis factor
    genes: structure homology and chromosomal
    localization. Nucleic Acids Res. 13:6361
    (1985).
8. M. A. Palladino Jr., M. R. Shalaby, S. M. Kramer et
    al. Characterization of the antitumour
    activities of human tumour necrosis factor alpha
    and the comparison with the other cytokines:

induction of tumour-specific immunity. J.
Immunol. 138:4023
(1987).

9. B. Beutler. Orchestration of septic shock by
   cytokines: The role of cachectin (tumour
   necrosis factor), in: "Molecular and Cellular
   Mechanisms of Septic Shock", Alan R. Liss, New
   York (1989).

10. B. Beutler, N. Krochin, I. W. Milsark, et al.
    Control of cachectin (tumour necrosis factor)
    synthesis: mechanisms of endotoxin resistance.
    Science 232:977 (1986).

11. D. G. Remick, R. M. Strieter, J. P. Lynch, et al.
    In vivo dynamics of murine tumour necrosis
    factor alpha gene expression. Kinetics of
    dexamethasone-induced suppression. Lab. Invest.
    60:766 (1989).

12. S. S. Boggs, D. R. Boggs, and R. A. Joyce.
    Response to endotoxin of endotoxin-resistant
    C3H/HeJ mice: a model for study of hematopoietic
    control. Blood 55:444 (1980).

13. K. J. Tracey, S. F. Lowry, and A. Cerami.
    Cachectin: a hormone that triggers acute shock
    and chronic cachexia. J. Infect. Dis. 157:413
    (1988).

14. S. Q. Simpson and L. C. Casey. Role of tumour
    necrosis factor in sepsis and acute lung injury.
    Critical Care Clinics 5:27 (1989).

15. M. Blick, S. A. Sherwin, M. Rosenblum, and J.
    Gutterman. Phase 1 study of recombinant tumour
    necrosis factor in cancer patients. Cancer Res.
    47:2986 (1987).

16. A. F. Suffredini, R. F. Fromm, M. M. Parker, et al.
    The cardiovascular response of normal humans to
    the administration of endotoxin. N. Engl. J.
    Med. 321:280 (1989).

17. H. R. Michie, K. R. Manogue, D. R. Spriggs, et al.
    Detection of circulating tumour necrosis factor
    after endotoxin administration. N. Engl. J. Med.
    318:1481 (1988).

18. B. A. Beutler, I. W. Milsark, and A. Cerami.
    Cachectin/tumour necrosis factor: production,
    distribution, and metabolic fate in vivo. J.
    Immunol. 135:3972 (1985).

19. B. Beutler, I. W. Milsark, and A. Cerami. Passive
    immunization against cachectin/tumour necrosis
    factor (TNF) protects mice from the lethal
    effect of endotoxin. Science 229:869 (1985).

20. K. J. Tracey, Y. Fong, D. G. Hesse, et al.
    Anti-cachectin/TNF monoclonal antibodies prevent

      septic shock during lethal bacteremia. Nature
      330:662 (1987).
21. C. A. Dinarello, and N. Savage. Interleukin-1 and
      its receptor. Critical Rev. Immunol. 9:1 (1989).
22. S. Demczuk, D. Baumberger, B. Mach, and J.-M.
      Dayer. Expression of human IL-1 alpha and beta
      messenger RNAs and IL-1 activity in human
      peripheral blood mononuclear cells. J. Mol.
      Cell. Immunol. 3:225 (1987).
23. B. Lepe-Zuniga, and I. Gery. Production of
      intracellular and extracellular interleukin-1
      (IL-1) by human monocytes. Clin. Immunol.
      Immunopathol. 31:222 (1984).
24. G. W. Duff and E. Atkins. The detection of
      endotoxin by in vitro production of endogenous
      pyrogen: Comparison with limulus amebocyte
      lysate gelation. J. Immunol. Methods. 52:323
      (1982).
25. A. Waage, P. Brandtzaeg, A. Halstensen, P. Kierulf,
      and T. Espevik. The complex pattern of
      cytokines in serum from patients with
      meningococcal septic shock. J. Exp. Med. 169:333
      (1989).
26. S. Okusawa, J. A. Gelfand, T. Ikejima, R. J.
      Connolly, and C. A. Dinarello. Interleukin 1
      induces a shock-like state in rabbits. Synergism
      with tumour necrosis factor and the effect of
      cyclooxygenase inhibition. J. Clin. Invest.
      81:1162 (1988).
27. C. A. Dinarello, J. G. Cannon, S. M. Wolff, et al.
      Tumour necrosis factor (cachectin) is an
      endogenous pyrogen and induces production of
      interleukin 1. J. Exp. Med. 163:1433 (1986).
28. P. J. Knudsen, C. A. Dinarello, and T. B. Strom.
      Glucocorticoids inhibit transcriptional and
      post-transcriptional expression of interleukin 1
      in U937 cells. J. Immunol. 139:4129 (1987).
29. J. R. Weinberg, D. J. M. Wright, and A. Guz.
      Interleukin-1 and tumour necrosis factor cause
      hypotension in the conscious rabbit. Clin.
      Science 75:251 (1988).
30. C. A. Dinarello. Interleukin-1. Rev. Infect. Dis.
      6:51 (1984).
31. S. E. Goldblum, D. A. Cohen, M. N. Gillespie and C.
      J. McClain. Interleukin-1-induced
      granulocytopenia and pulmonary leukostasis in
      rabbits. J. Appl. Physiol. (in press).
32. C. A. Dinarello, S. Okusawa, and J. A. Gelfand.
      Interleukin-1 induces a shock-like state in
      rabbits: synergism with tumour necrosis factor
      and the effect of cyclooxygenase inhibition. in:

Molecular and Cellular Mechanisms of Septic
Shock. Alan R. Liss, New York (1989).

33. B. Everaerdt, P. Brouckaert, A. Shaw, and W. Fiers.
Four different interleukin-1 species sensitize
to the lethal action of tumour necrosis factor.
Biochem. Biophys. Res. Comm. 163:378 (1989).

34. R. M. Kramer, J. A. Jakubowski, and D. Deykin.
Hydrolysis of 1-alkyl-2-arachidonoyl-sn-glycero-
3-phosphocholine, a common precursor of
platelet-activating factor and eicosanoids, by
human platelet phospholipase A2. Biochim.
Biophys. Acta 959:269 (1988).

35. F. H. Chilton, M. J. Ellis, S. C. Olson, and R. L.
Wykle. 1-O-alkyl-2-arachidonyl-sn-glycero-3-
phosphocholine: a common source of platelet
activating factor and arachidonate in human
polymorphonuclear leukocytes. J. Biol. Chem.
259:12014 (1984).

36. T. Ch. Lee, B. Malone, and F. Snyder. A new de
novo pathway for the formation of
1-alkyl-2-acetyl-sn-glycerols, precursors of
platelet activating factor. Biochemical
characterization of 1-alkyl-2-lyso-sn-
glycero-3-P: acetyl-CoA acetytransferase in rat
spleen. J. Biol. Chem. 261:5373 (1986).

37. P. Braquet and M. Rola-Pleszcynski.
Platelet-activating factor and cellular immune
responses. Immunol. Today 8:345 (1987).

38. Y. Thomas, Y. Denizot, E. Dassa, C. Boullet, and J.
Benveniste. Synthese du paf-acether par E. coli
K12. C. R. Acad. Sci. Paris 303:699 (1986).

39. P. Braquet, M. Paubert-Braquet, P. Bessin, and B.
B. Vargaftig. Platelet-activating factor: a
potential mediator of shock. Adv. Prostaglandin
Thromboxane Leukotriene Res. 17:822 (1987).

40. X.-M. Sun and W. Hsueh. Bowel necrosis induced by
tumour necrosis factor in rats is mediated by
platelet-activating factor. J. Clin. Invest.
81:1328 (1988).

41. K.-M. Chu, J. G. Gerber and A. S. Nies. Local
vasodilator effect of platelet activating factor
in the gastric, mesenteric and femoral arteries
of the dog. J. Pharmacol. Exp. Ther. 246:996
(1988).

42. V. Lagente, Z. B. Fortes, J. Garcia-Leme, and B. B.
Vargaftig. PAF-acether and endotoxin display
similar effects on rat mesenteric microvessels:
inhibition by specific antagonists. J.
Pharmacol. Exp. Ther. 247:254 (1988).

43. G. Feuerstein, and A. L. Siren.

Platelet-activating factor and shock. Prog.
Biochem. Pharmacol. 22:181 (1988).

44. C. Kroegel. The potential pathophysiological role
of platelet-activating factor in human diseases.
Klin. Wochenschr. 66:373 (1988).

45. A. M. Lefer. Eicosanoids as mediators of ischemia
and shock. Federation Proc. 44:275 (1985).

46. A. M. Lefer. Leukotrienes as mediators of ischemia
and shock. Biochem. Pharmacol. 35:123 (1986).

47. W. Oettinger, B. A. Peskar, and H. G. Beger.
Profiles of endogenous prostaglandins F2a,
thromboxane A2 and prostacyclin with regard to
cardiovascular and organ functions in early
septic shock in man. Eur. Surg. Res. 19:65
(1987).

48. M. E. Doerfler, R. L. Danner, J. H. Shelhamer, and
J. E. Parrillo. Bacterial lipopolysaccharides
prime human neutrophils for enhanced production
of leukotriene B4. J. Clin. Invest. 83:970
(1989).

49. G. Feuerstein, and J. M. Hallenbeck.
Prostaglandins, leukotrienes, and
platelet-activating factor in shock. Annu. Rev.
Pharmacol. Toxicol. 27:301 (1987).

50. T. Schonhardt and E. Ferber. Translocation of
phospholipase A2 from cytosol to membranes
induced by 1-oleoyl-2-acetyl-glycerol in
serum-free cultured macrophages. Biochem.
Biophys. Res. Comm. 149:769 (1987).

51. W. Pruzanski, P. Vadas, J. Kim, H. Jacobs, and E.
Stefanski. Phospholipase A2 activity associated
with synovial fluid cells. J. Rheumatol. 15:791
(1988).

52. J. Seilhamer, P. Vadas, W. Pruzanski, S. Plant, E.
Stefanski and L. Johnson. Synovial fluid
phospholipase A2 in arthritis. in: "Therapeutic
approaches to inflammatory diseases". A. J.
Lewis, N. S. Doherty, and N. R. Ackerman, eds.,
Elsevier, New York (1989).

53. J. J. Seilhamer, W. Pruzanski, P. Vadas, S. Plant,
J. A. Miller, J. Kloss, and L. K. Johnson.
Cloning and recombinant expression of
phospholipase A2 present in rheumatoid arthritic
synovial fluid. J. Biol. Chem. 264:5335 (1989).

54. J. J. Seilhamer, S. Plant, W. Pruzanski, J.
Schilling, E. Stefanski, P. Vadas, and L. K.
Johnson. Multiple forms of phospholipase A2 in
arthritic synovial fluid. J. Biochem. 106:38
(1989).

55. R. M. Kramer, C. Hession, B. Johansen, et al.

Structure and properties of a human non-pancreatic phospholipase A2. J. Biol. Chem. 264:5768 (1989).

56. C.-Y. Lai and K. Wada. Phospholipase A2 from human synovial fluid: purification and structural homology to the placental enzyme. Biochem. Biophys. Res. Comm. 157:488 (1988).

57. P. Vadas, W. Pruzanski, E. Stefanski, et al. Extracellular phospholipase A2 secretion is a common effector pathway of interleukin-1 and tumour necrosis factor action. (submitted for publication).

58. J. Chang, S. C. Gilman, and A. J. Lewis. Interleukin 1 activates phospholipase A2 in rabbit chondrocytes: a possible signal for IL-1 action. J. Immunol. 136:1283 (1986).

59. S. C. Gilman, J. Chang, P. R. Zeigler, J. Uhl, and E. Mochan. Interleukin-1 activates phospholipase A2 in human synovial cells. Arthritis Rheumatism 31:126 (1988).

60. J. Pfeilschifter, W. Pignat, K. Vosbeck, and F. Marki. Interleukin 1 and tumour necrosis factor synergistically stimulate prostaglandin synthesis and phospholipase A2 release from rat renal mesangial cells. Biochem. Biophys. Res. Comm. 159:385 (1989).

61. C. Lanni and E. Becker. Release of phospholipase A2 from rabbit peritoneal neutrophils by f-Met-Leu-Phe. Am. J. Pathol. 113:90 (1983).

62. P. Vadas and J. B. Hay. The release of phospholipase A2 from aggregated platelets and stimulated macrophages of sheep. Life Sci. 26:1721 (1980).

63. P. Vadas, W. Pruzanski, E. Stefanski, et al. Pathogenesis of hypotension in septic shock: correlation of circulating phospholipase A2 levels with circulatory collapse. Critical Care Med. 16:1 (1988).

64. P. Vadas and J. B. Hay. The appearance and significance of phospholipase A2 in lymph draining tuberculin reactions. Am. J. Pathol. 107:285 (1982).

65. J. R. Traynor and K. S. Authi. Phospholipase A2 activity of lysosomal origin secreted by polymorphonuclear leukocytes during phagocytosis or on treatment with calcium. Biochim. Biophys. Acta 665:571 (1981).

66. P. Vadas and J. B. Hay. Involvement of circulating phospholipase A2 in the pathogenesis of the hemodynamic changes in endotoxin shock in

rabbits. Can. J. Physiol. Pharmacol. 61:561
(1983).

67. P. Vadas, S. Wasi, H. Z. Movat and J. B. Hay.
Extracellular phospholipase A2 mediates
inflammatory hyperemia. Nature 273:583 (1981).

68. H.-C. Huang. Release of slow reacting substance
from the guinea-pig lung by phospholipases A2 of
Vipera Russelli snake venom. Toxicon 22:359
(1984).

69. J. Shaw, M. Roberts, R. Ulevitch, P. Henson, and E.
Dennis. Phospholipase A2 contamination of cobra
venom factor preparations. Am. J. Pathol. 91:571
(1978).

70. J. Edelson, W. Pruzanski, E. Stefanski and P.
Vadas. Type I venom phospholipase A2 induces
acute lung injury resembling ARDS. (submitted
for publication).

71. P. Vadas, W. Pruzanski, J. Kim and V. Fornasier.
The proinflammatory effect of intra-articular
injection
of soluble human and venom phospholipase A2. Am.
J. Pathol. 134:807 (1989).

72. B. S. Vishwanath, A. A. Fawzy, and R. C. Franson.
Edema-inducing activity of phospholipase A2
purified from human synovial fluid and
inhibition by aristolochic acid. Inflammation
12:549 (1988).

73. L. A. Marshall, J. Y. Chang, W. Calhoun, J. Yu, and
R. P. Carlson. Preliminary studies on
phospholipase A2-induced mouse paw edema as a
model to evaluate antiinflammatory agents. J.
Cell. Biochem. 40:147 (1989).

74.   G. Cirino, S. H. Peers, J. L. Wallace, and R. J.
Flower.   A study of phospholipase A2-induced
paw edema. Eur. J. Pharmacol. 166:505 (1989).

75. G. E. Hoffmann, R. Hiefinger, and B.
Steinbrueckner. Serum phospholipase A in
hospitalized patients. Clin. Chim. Acta 183:59
(1989).

76. P. Vadas. Elevated plasma phospholipase A2 levels:
correlation with the hemodynamic and pulmonary
changes in gram-negative septic shock. J. Lab.
Clin. Med. 104:873 (1984).

77. G. Offenstadt, P. Pinta, J. Masliah, et al.
Phospholipasic and prophospholipasic activities
in bronchoalveolar lavage fluid in severe acute
pulmonary disease with or without ARDS.
Intensive Care Med. 7:285 (1981).

78. P. Vadas, E. Stefanski and W. Pruzanski. Influence
of plasma proteins on activity of

pro-inflammatory enzyme phospholipase A2. Inflammation. 10:183 (1986).

79. K. M. Conricode and R. S. Ochs. Mechanism for the inhibitory and stimulatory actions of proteins on the activity of phospholipase A2. Biochim. Biophys. Acta. 1003:36 (1989).

80. P. Vadas, W. Pruzanski, E. Stefanski, et al. Concordance of endogenous cortisol and phospholipase A2 levels in gram-negative septic shock. A prospective study. J. Lab. Clin. Med. 111:584 (1988).

81. W. Pruzanski, V. Farewell, and P. Vadas. Kinetics of phospholipase A2 clearance during convalescence from septic shock. (manuscript in preparation).

82. S. C. Gilman. Activation of rabbit articular chondrocytes by recombinant human cytokines. J. Rheumatol. 14:1002 (1987).

83. W. Pruzanski and P. Vadas. Studies on proinflammatory phospholipase A2, and the cytokine effector network. in: Biochemistry, molecular biology and physiology of phospholipase A2 and its regulatory factors. A. B. Mukherjee ed., Plenum Press, New York (in press).

84. W. Pruzanski, D. Wilmore, E. Stefanski and P. Vadas. Endotoxin-induced intravascular secretion of phospholipase A2 in healthy human volunteers: Relationship to tumour necrosis factor. (manuscript in preparation).

85. E. Ziegler, J. McCutchan, J. Fierer, et al. Treatment of gram-negative bacteremia and shock with human antiserum to a mutant E. coli. N. Engl. J. Med. 307:1226 (1982).

MOBILIZATION AND FUNCTION OF EXTRACELLULAR PHOSPHOLIPASE A$_2$ IN INFLAMMATION[1]

Jerrold Weiss and Grace Wright

New York University School of Medicine

Departments of Microbiology and Medicine, NY, NY

A common characteristic of inflammatory exudates is the accumulation of relatively high concentrations of phospholipase A$_2$ (PLA-2) in the cell-free fluid (1). Because many (by-)products of this enzyme's action can exert pro-inflammatory effects (e.g. eicosanoid derivatives of arachidonic acid), these observations have prompted considerable speculation about the possible (patho-) physiological role of extracellular PLA-2 in inflammation. These ideas, in turn, have triggered many investigators and pharmaceutical companies alike to search for "specific" inhibitors of PLA-2 that could be applied therapeutically in acute or chronic inflammatory diseases.

Despite these efforts, many fundamental questions remain that need to be addressed in order to obtain a clearer understanding of the significance of extracellular PLA-2 in inflammation and of the possible means to specifically modulate its action in vivo. These questions include the source(s) of extracellular PLA-2 in inflammation, the multiplicity and diversity of these enzymes, their molecular properties and biological target(s). All extracellular PLA-2 characterized to date (from insect to man) exhibit a common set of structural and functional attributes indicative of a family of highly conserved enzymes(2). These closely similar proteins can, nevertheless, differ considerably in their action on the phospholipids of natural membranes, possibly reflecting differences in "(hyper-)variable" regions that may determine ancillary functions needed for enzyme action on specific biological targets (3,4). A single organism can elaborate multiple extracellular PLA-2, presumably to perform different biological functions (2-4). The

---

[1]This work was supported by USPHS Grants AI 18571 and R37 DK 05472.

Phospholipase A2
Edited by P.Y.-K. Wong and E. A. Dennis
Plenum Press, New York, 1990

mapping of variable regions may ultimately facilitate decoding of
PLA-2 function and also provide a target for truly specific inhibitors of PLA-2 function.

     Our studies on the mobilization and function of extracellular
PLA-2 in inflammation have evolved from our interest in the biochemical mechanisms of host defense against infection. The defense
of the host from invading microorganisms is generally assumed to
require ultimaly digestion of microbial constituents but experimental evidence of the participants and determinants of this
process is scanty. Our focus has been on the determinants of
bacterial phospholipid degradation which accompanies the antibacterial action of polymorphonuclear leukocytes (PMN), first-line
cells in host defense, since phospholipids are essential structural
elements of cells and thus their breakdown might reflect and help
determine the overall destruction of the bacterium. Using
Escherichia coli as a test microorganism, we have found that the
antibacterial action of PMN is accompanied by bacterial phospholipid degradation in which the bacterial outer membrane
phospholipase A and both a PMN and an inflammatory fluid PLA-2
participate (5,6). It has been possible to establish the
participation of each of these PLA by making use of mutant strains
of E. coli that vary in their expression of the pldA gene encoding
the bacterial outer membrane PLA (7,8). Because the membrane
phospholipids of E. coli, like that of other cells, is normally
resistant to the action of either endogenous or exogenous
phospholipases (5,9), PLA action on bacteria ingested by PMN
 implies the participation of additional factor(s) which create
conditions permissive for PLA activity. We have isolated a
membrane-active bactericidal protein (bactericidal/permeability-
increasing protein (BPI)) from PMN which activates bacterial
phospholipid degradation by both the bacterial PLA and the host
PLA-2 in a manner remarkably similar to that produced when E. coli
are ingested by intact PMN (5,9).

PLA-2 ACTION ON BPI-TREATED E. COLI IS HIGHLY SELECTIVE

     BPI treatment of E. coli triggers bacterial phospholipid
degradation by only certain PLA-2 suggesting that this is a
functional attribute dependent on some variable "domain" of PLA-2
(5). Chemical modification and primary structure analyses of a so-
called "BPI-responsive" enzyme that was available in sufficient
quantities for these structural studies (the basic isozyme of
agkistrodon halys blomhoffii venom) revealed a cluster of basic
residues in the NH_2-terminal 15 residues that appeared to be important for enzyme activity toward BPI-treated E. coli but not
other substrates including autoclaved E. coli (10,11). In all
PLA-2 sequenced to date, residues 1-12(13) are predicted (12)
(or have been observe;13) to form an alpha-helix. An axial

MOBILIZATION AND FUNCTION IN INFLAMMATION 105

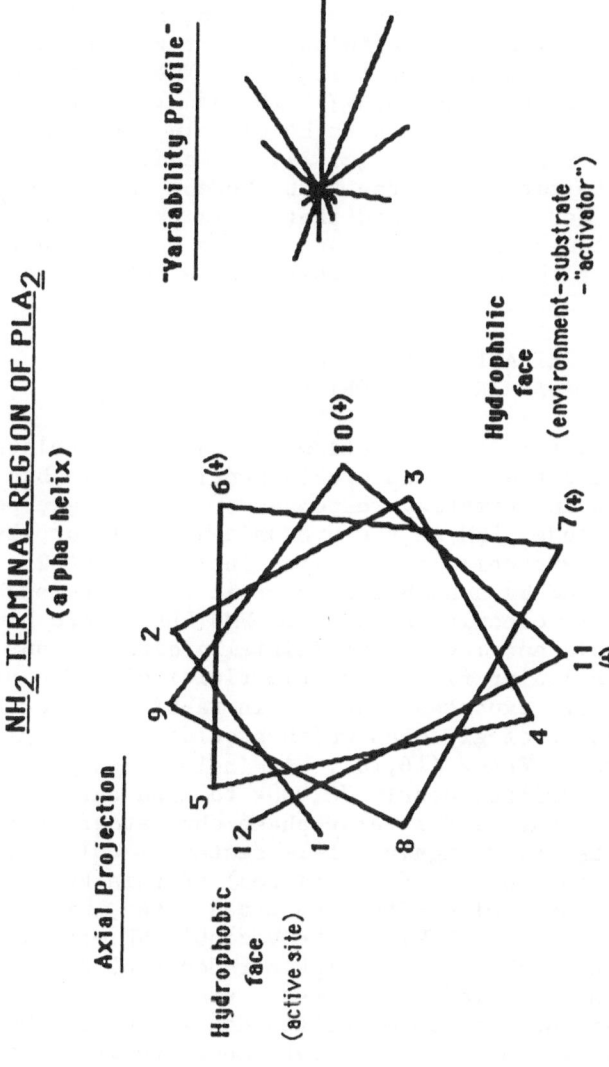

Fig. 1. Schematic of: a) spatial arrangement of PLA-2 NH₂-terminal region and surroundings; b) "variability profile" of individual residues within this region. The length of the line is proportional to the degree of variation of that residue in PLA-2 (14). (+) denotes site of basic residue in PLA-2 active against BPI-treated E. coli.

projection of this region reveals an asymmetric arrangement of polar
and nonpolar amino acids that is common to all PLA-2, providing an
amphipathic structure that is probably essential for enzyme-
substrate interaction at a lipid-H_2O (membrane) interface. The
nonpolar face is thought to form a hydrophobic wall in the active
site cavity that is nearly invariant in composition (15; Fig. 1).
The polar face, in contrast, is directed outward to the environment and its composition varies greatly, creating a "variable
surface domain" that may determine (in part) differences in activity
among PLA-2 toward specific targets (5,11,15). The basic agkistrodon h.b. PLA-2 contains Arg-6 and Lys-7, -10 and -11 that,
according to our model are predicted to be closely juxtaposed along
the polar face of the alpha-helix (Fig. 1). In addition, residues
13-16 are predicted to form a reverse turn thereby placing Lys-15
in close proximity. We have proposed that it is this unusual
cluster of basic residues in this surface region that accounts
(at least in part) for the ability of this enzyme but not most
other PLA-2 to act on BPI-treated E. coli.

ISOLATION AND CHARACTERIZATION OF A PLA-2 FROM THE
EXTRACELLULAR FLUID OF AN INFLAMMATORY EXUDATE

    How do these structural and functional predictions relate to
the properties of PLA-2 that may naturally participate in bacterial
digestion? The acute inflammatory response normally triggered by
invading bacteria includes PMN as the predominant cellular element
and a complex mix of extracellular proteins including PLA-2. This
response can be mimicked by a number of sterile irritants such as
glycogen(16). Because molecular studies of PMN PLA-2 were hampered
by the scarcity of this enzyme(17), we initially directed our
efforts at the extracellular fluid (ascitic fluid) of PMN-contain-
ing sterile inflammatory exudates elicited in rabbits following
intraperitoneal infusion of glycogen which provides a much richer
source of "inflammatory" PLA-2 (16,18). A single, highly basic
(pI 10.5), PLA-2 was purified nearly 70,000x to apparent
homogeneity by ion-exchange and reverse-phase chromatographies (18).
The purified enzyme is active against BPI-treated E. coli. The
NH_2-terminal amino acid sequence (39 residues) of this PLA-2 shows
a high degree of homology, within the conserved sites, with all
PLA-2 and extensive homology (75%) in the variable NH_2-terminal
region (residues 1-16) with the basic agkistrodon h.b. PLA-2 (Fig.
2). Most notably, the ascitic fluid PLA-2 contains a cluster of
basic residues within the "variable surface domain" of the NH_2-
terminal region (Fig. 1) that is, thus far, characteristic of PLA-2
active against BPI-treated E. coli. This remarkable degree of
homology between functionally similar enzymes from widely divergent
sources lends further support to our concept that spatially apposed
basic amino acids in the NH_2-terminal region are important determinants of PLA-2 action toward BPI-treated E. coli and possibly

PLA-2	Activity vs BPI-ΔE. coli	1					6	7			10	11				15
Basic a.h.b.	+	His	Leu	Leu	Gln	Phe	ARG	LYS	Met	Ile	LYS	LYS	Met	Thr	Gly	LYS
Rabbit Ascitic fl.	+	His	Leu	Leu	Asp	Phe	ARG	LYS	Met	Ile	ARG	Tyr	Thr	Thr	Gly	LYS
Rabbit PMN	+	Ala	Leu	Leu	Asp	Phe	ARG	LYS	Met	Ile	ARG	Tyr	Thr	Thr	Gly	LYS
Rabbit Serum	+	His	Leu	Leu	Asp	Phe	ARG	LYS	Met	Ile	ARG	Tyr	Thr	Thr	Gly	LYS
Rabbit Platelet (secretory)	?	His	Leu	Leu	Asp	Phe	ARG	LYS	Met	Ile	ARG	Tyr	Thr	Thr	Gly	LYS
Human Synovial fl.	+	Asn	Leu	Val	Asn	Phe	His	ARG	Met	Ile	LYS	Leu	Thr	Thr	Gly	LYS
Pig Intestinal	+	Asp	Leu	Leu	Asn	Phe	ARG	LYS	Met	Ile	LYS	Leu	Lys	Thr	Gly	LYS
Pig Pancreas	−	Ala	Leu	Trp	Gln	Phe	ARG	Ser	Met	Ile	LYS	Cys	Ala	Ile	Pro	Gly

Figure 2   NH$_2$-TERMINAL AMINO ACID SEQUENCES OF BPI-RESPONSIVE AND -NONRESPONSIVE PLA-2

characteristic of PLA-2 participating in bacterial digestion.

Amino acid sequence analysis also revealed that the ascitic fluid PLA-2 lacks Cys-11 and contains a short C-terminal extension terminating with a Cys, hallmarks of "group II" PLA-2 (19). This was the first definitive demonstration that mammalian species contain a representative of group II PLA-2; pancreatic PLA-2 belong to group I (19). Several other examples of mammalian group II PLA-2 (e.g. platelet, synovial fluid, intestinal, PMN, serum; 20-23) have since been reported. With the exception of the human pancreatic PLA-2 which exhibits ca. 1-2% of the activity of the ascitic fluid PLA-2 toward BPI-treated E. coli (24), all currently recognized "BPI-responsive" PLA-2 belong to group II although not all group II PLA-2 are "BPI-responsive."

Our studies have not revealed any evidence of multiplicity or diversity of PLA-2 in the cell-free inflammatory fluid. It is possible, however, that other PLA-2 have gone unnoticed during purification of the ascitic fluid PLA-2 because of our exclusive use of autoclaved E. coli as substrate. Although virtually all (>25) PLA-2 that we have tested display nearly the same specific activity toward this substrate, other PLA-2 activities have been reported which show a more restricted substrate preference (25).

CONTRIBUTION OF ASCITIC FLUID PLA-2 TO DIGESTION OF E. COLI INGESTED BY PMN

Does the presence in acute inflammatory exudates in which PMN accumulate of relatively large amounts of an extracellular PLA-2 that is structurally and functionally equipped to act against BPI-treated E. coli imply a role for this enzyme in bacterial digestion by the host?

We have approached this question by comparing phospholipid degradation in the PLA-less (pldA$^-$) strain of E. coli during incubation with PMN in the presence and absence of added ascitic fluid PLA_2. The added PLA-2 stimulated bacterial phospholipid degradation as much as 4-fold (from ca. 7 to 30%) during 60 min incubations. The effect of the purified enzyme was nearly identical to that of the whole ascitic fluid indicating that no other factors in the inflammatory exudate were required for the enhanced phospholipid degradation (6).

In these experiments, bacterial uptake by PMN was complete within 15-30 min yet phospholipid degradation continued for at least 60 min suggesting that the added extracellular PLA-2 was acting on intracellular bacteria. Ascitic fluid PLA-2 added after all E. coli had been ingested (≥30 min) did not increase bacterial phospholipid degradation, confirming that the PMN-associated bacteria were indeed intracellular and suggesting that the action of the ascitic fluid PLA-2 may require co-internalization of the enzyme with the bacteria.

Relatively large amounts of the ascitic fluid PLA-2 bind rapidly to both E. coli and to PMN. PLA-2 binding was the same when either whole ascitic fluid or purified PLA-2 was added indicating that no other exudate proteins were required to promote PLA-2 binding. Binding of purified PLA-2 to either E. coli or to PMN did not cause phopholipid degradation in either cell. Nevertheless, precoating of either bacteria or PMN with PLA-2 enhanced bacterial phospholipid degradation during phagocytosis by PMN in a manner similar to that seen when the PLA-2 was added directly to the incubation mixture. Thus, the extracellular PLA-2 can apparently enter the phagocytic vacuole either "piggy-back" on the E. coli or on invaginated PMN plasma membrane and, once inside, can contribute to the hydrolysis of the bacterial phospholipids.

WHERE IS THE ORIGIN OF THE ASCITIC FLUID PLA-2?

The cellular source(s) of extracellular PLA-2 during inflammation remain(s) uncertain. Available evidence indicates that the predominant PLA-2 in synovial fluid of patients with rheumatoid arthritis is indistinguishable from a granule-associated (secretory) enzyme present in human platelets (26). However, in rabbit, where more extensive primary structure is known, discrete differences have been found between inflammatory (ascitic) fluid and platelet PLA-2 (27), indicating that the two enzymes are closely similar but distinct.

Stimulated (rabbit peritoneal exudate) PMN also secrete PLA-2 during degranulation in vitro (28) suggesting that these cells may be a source of inflammatory fluid PLA-2. However, the total cellular content of PLA-2 and the amount released represent barely a trace of the activity present in ascitic fluid containing a similar number of PMN (28,29), all measured using autoclaved E. coli as substrate. Moreover, primary structure analysis of a granule-associated PLA-2 that we have recently purified from rabbit PMN shows that the PMN and ascitic fluid PLA-2 are identical at 15 of 16 residues in the NH_2-terminal region but differ at the NH_2-terminus (29; Fig. 2) indicating that these two PLA-2 are also closely similar but distinct.

We have also isolated a PLA-2 from rabbit serum and this enzyme shares all of the first 19 residues with the ascitic fluid PLA-2 (29; Fig. 2) suggesting that the ascitic fluid enzyme may originate in the plasma during formation of the inflammatory exudate. The level of PLA-2 activity in the ascitic fluid reaches levels that are comparable to those in serum or plasma over a time course that seems consistent with transudation of this enzyme from plasma.

Figure 3. Putative locations of predicted neurotoxic (98–111),
myotoxic (88–98), anti-coagulant (54–77) and "BPI-responsive"
(6–15) sites of PLA-2 (adapted from Kini, M.R. and Evans, H.J.
J. Biol. Chem. 262:14402, 1987). Polypeptide backbone is from
X-ray structure of bovine pancreas PLA-2.

FUTURE PERSPECTIVES

The studies briefly described above provide the first experimental evidence of a possible physiological function for extracellular PLA-2 in inflammatory exudates. The presence of structurally and functionally very closely similar PLA-2 in the cellular and extracellular compartments of an inflammatory exudate is consistent with the apparent role of these enzymes in the destruction of certain microbial invaders during the acute inflammatory response. Similar structural and functional attributes have been also observed in a human inflammatory fluid PLA-2 isolated from synovial fluid and donated to us by Dr. Tom Parks and, in addition, in a porcine intestinal PLA-2 purified from Paneth cells (22) that may also have bacteria as a natural target (Fig. 2).

The selective activation by BPI of PLA-2 acting on the phospholipids of bacteria killed by BPI has provided us with a unique biological setting in which to explore what determines functional differences among these highly conserved enzymes. This had led to the identification of variables in the NH_2-terminal region of PLA-2 that apparently (co-) determine differences in the ability of various enzymes to act on this biological target. Structural variability in other (surface) regions of the PLA-2 molecule has been related by other investigators to differences among these enzymes in other biological actions (4; Fig. 3). Taken together, these models portray the surface of a PLA-2 as a mosaic of variable domains that may underly the functional diversification of this enzyme family and also may mean that any single PLA-2 can have multiple biological targets and functions. Cloned PLA-2 genes, including that encoding the synovial fluid enzyme (24,26), that can be expressed and discretely mutated should greatly facilitate the dissection of the structural determinants of PLA-2 action vs. BPI-killed E. coli and of PLA-2 functional diversity in general.

REFERENCES

1. P. Vadas and W. Pruzanski, Role of extracellular phospholipase A_2 in inflammation, Adv. Inflamm. Res. 7:51 (1984).
2. M. Waite, "The Phospholipases," Plenum Press, New York (1987).
3. P. Elsbach, J. Weiss, and S. Forst, Determinants of the action of phospholipases on the phospholipids of gram-negative bacteria (E. coli). In: Lipids and Biomembranes: Past, Present and Future. J. Op Den Kamp, B. Roelofsen and K.W.A. Wirtz, eds., pp. 259-286, Elsevier Science Publishers, Amsterdam, The Netherlands (1986).
4. R.M. Kini and H.J. Evans, Structure-function relationships of phospholipases. The anticoagulant region of phospholipase A_2, J. Biol. Chem. 262:14402 (1987).

5. P. Elsbach and J. Weiss, Phagocytosis of bacteria and phospholipid degradation, Biochim. Biophys. Acta (Reviews of Biomembranes)
947:29 (1988).

6. G. Wright, J. Weiss and P. Elsbach, Role of extracellular and
cellular phospholipases in the destruction of E. coli in an
inflammatory exudate, Clin Res 36:475A (1988).

7. O. Doi and S. Nojima, Nature of Escherichia coli mutants
deficient in detergent-resistant and/or detergent-sensitive
phospholipase A, J. Biochem. 80:1247 (1976).

8. P. de Geus, I. van Die, H. Bergmans, J. Tommassen and G.H. de
Haas, Molecular cloning of pldA, the structural gene for outer
membrane phospholipase of E. coli K12, Mol. Gen. Genet. 190:150
(1983).

9. J. Weiss, S. Beckerdite-Quagliata and P. Elsbach, Determinants
of the action of phospholipases A on the envelope phospholipids
of Escherichia coli, J. Biol. Chem. 254:11010 (1979).

10. S. Forst, J. Weiss, and P. Elsbach, The role of phospholipase
A_2 lysines in phospholipolysis of Escherichia coli killed by a
membrane-active neutrophil protein, J. Biol. Chem. 257:14055 (1982).

11. S. Forst, J. Weiss, P. Blackburn, B. Frangione, F. Goni and
P. Elsbach, Amino acid sequence of a basic Agkistrodon halys
blomhoffii phospholipase A_2: Possible role of NH_2-terminal lysines
in action on phospholipids of Escherichia coli, Biochemistry 25:
4309 (1986).

12. M.J. Dufton, D. Eaker and R.C. Hider, Conformational properties of phospholipases A_2: Secondary structure prediction,
circular dichroism and relative interface hydrophobicity,
Eur. J. Biochem. 137:537 (1983).

13. R. Renetseder, S. Brunie, B.W. Dijkstra, J. Drenth and P.B.
Sigler, A comparison of the crystal structures of phospholipase
A_2 from bovine pancreas and Crotalus Atrox venom, J. Biol. Chem.
260:11627 (1985).

14. G.E. Schulz and R.H. Schirmer, "Principles of protein structure"
Springer-Verlag, New York (1979).

15. A. Randolph and R.L. Heinrikson, Crotalus atrox phospholipase
A_2. Amino acid sequence and studies on the function of the NH_2-
terminal region, J. Biol. Chem. 257:2155 (1982).

16. R. Franson, R. Dobrow, J. Weiss, P. Elsbach and W. Weglicki,
Isolation and characterization of a phospholipase A_2 from an
inflammatory exudate, J. Lipid Res. 19:18 (1978).

17. P. Elsbach, J. Weiss, R.C. Franson, S. Beckerdite-Quagliata,
A. Schneider and L. Harris, Separation and purification of a
potent bactericidal/permeability-increasing protein and a closely
associated phospholipase A_2 from rabbit polymorphonuclear leukocytes, J. Biol. Chem. 254:11000 (1979).

18. S. Forst, J. Weiss, P. Elsbach, J.M. Maraganore, I. Reardon
and R.L. Heinrikson, Structural and functional properties of a
phospholipase A_2 purified from an inflammatory exudate, Biochemistry 25:8381 (1986).

19.   R.L. Heinrikson, E.T. Krueger, and P.S. Keim, Amino acid sequence of phospholipase A_2-alpha from the venom of Crotalus adamanteus: A new classification of phospholipase A_2 based upon structural determinants, J.Biol. Chem. 252:4913 (1977).

20.   H. Mizushima, I. Kudo, K. Horigome, M. Murakami, M. Hayakawa, D.K. Kim, E. Kondo, M. Tomita and K. Inoue, Purification of rabbit platelet secretory phospholipase A_2 and its characteristics, J. Biochem. 105:520 (1989).

21.   R.M. Kramer, C. Hession, B. Johansen, G. Hayes, P. McGray, E.P. Chow, R. Tizard and R.B. Pepinsky, Structure and properties of a human non-pancreatic phospholipase A_2, J. Biol. Chem. 264: 5768 (1989).

22.   R. Verger, F. Ferrato, C.M. Mansbach and G. Pieroni, A novel intestinal phospholipase A_2: Purification and some molecular characteristics, Biochemistry 21:6883 (1982).

23.   G.C. Wright, C.E. Ooi, J. Weiss and P. Elsbach, Purification of a cellular (granulocyte) and an extracellular (serum) phospholipase A_2 that participate in the destruction of Escherichia coli in a rabbit inflammatory exudate, submitted for publication, (1989).

24.   G. Wright, J. Weiss, J. Van den Bergh, H. Verheij and P. Elsbach, Genetic engineering of pig pancreas phospholipase A_2 to convert an inactive to an active ezyme in the bactericidal/ permeability-increasing protein (BPI)-mediated bacterial phospho- lipolysis, Clin. Res. 37:444A (1989).

25.   D.K. Kim, I. Kudo and K. Inoue, Detection in human platelets of phospholipase A_2 activity which preferentially hydrolyzes an arachidonoyl residue, J. Biochem. 104:492 (1988).

26.   J.J. Seilhamer, W. Pruzanski, P. Vadas, S. Plant, J.A. Miller, J. Kloss and L.K. Johnson, Cloning and recombinant expression of phospholipase A_2 present in rheumatoid arthritic synovial fluid, J. Biol. Chem. 264:5335 (1989).

27.   See reference # 20.

28.   C. Lanni and E.L. Becker, Release of phospholipase A_2 activity from rabbit peritoneal neutrophils by f-Met-Leu-Phe, Am. J. Pathol. 113:90 (1983).

29.   See reference # 23.

G-PROTEINS AND PHOSPHOLIPASE ACTIVATION IN ENDOTHELIAL CELLS

Mary E. Gerritsen and Robert J. Mannix

Department of Physiology
New York Medical College
Valhalla, N.Y.

SUMMARY

ATP stimulates arachidonic acid release and prostaglandin biosynthesis (most likely via phospholipase A_2 (PLA_2) activation) and phospholipase C (PLC) activation in cultured rabbit coronary microvessel endothelial cells. Pertussis toxin pretreatment inhibits ATP stimulated prostaglandin release, but not ATP stimulated phosphatidylinositol turnover. In contrast, activation of G-proteins with GTPγS or AlF_4- stimulates both prostaglandin synthesis and PLC. These observations suggest that PLC activation by ATP involves a G-protein(s) that is not ADP-ribosylated by pertussis toxin and further, that ATP activation of prostaglandin biosynthesis appears to involve a different, pertussis toxin sensitive, G-protein.

INTRODUCTION

G-proteins have been implicated in the coupling of membrane receptors to adenylate cyclase, PLC, certain ion channels and PLA_2 (1,2). In the present study, we report on our recent observations on the possible role of G-proteins in ATP stimulated prostaglandin release and PLC activation in microvessel endothelial cells.

METHODS

Materials

α-myo[2-^{3}H]inositol, [^{3}H]PGE$_2$ and [^{3}H]6-keto PGF$_\alpha$ were from

Amersham (Arlington Heights, Il). ATP, NaF and AlCl₃ were from Sigma (St. Louis, Mo.). All cell culture materials were from Gibco(Grand Island, NY). Pertussis toxin was from List Biologicals (Campbell,CA), Guanosine 5'-0-(3-thio)triphosphate (GTPγS) was from Boehringer Mannheim (Mannheim, FRG). Dowex 1-X8, formate form was from Biorad (Richmond, CA) and prostaglandin antisera were from Advanced Magnetics (Cambridge, MA).

Cell Culture

Endothelial cells used in this study were from an established cell line (RCME) derived from rabbit coronary microvessels. Details on the culture, identification and other characteristics of these cells have been provided in earlier publications (3,4).

Prostaglandin Radioimmunoassay

Prostaglandin E_2(PGE$_2$) and 6-keto prostaglandin F_α (6-keto PGF$_\alpha$, the stable hydrolysis product of PGI$_2$) levels in cell incubation buffers were determined by radioimmunoassay as previously described (3-5). Studies of prostaglandin production were routinely performed as follows. RCME cells were cultured on 24-well Falcon plates. Cells were used for the studies described below within 2-3 days of attaining confluence. Forty-eight hours prior to testing, cells were fed with fresh culture media (Dulbecco's Modified Eagles's Medium + 20% fetal bovine serum). The day of the experiment, culture media were aspirated and the cells washed gently with 3 x 1 ml of Dulbecco's phosphate buffered saline containing Ca⁺⁺ and Mg⁺⁺ (PBS). All buffers, drugs etc. were pre-equilibrated to 37°C and all incubations were carried out at 37°C. Cells were pre-incubated 15 min in PBS, the preincubation buffer removed, and PBS containing drugs or agonist added and incubated for the indicated period of time. Since basal and stimulated prostaglandin release varied somewhat from one multiwell plate to another, every multi-well plate contained negative control (PBS or other buffer) and positive control (500μM)ATP groups. Within each experiment each group had an n=4-6 and all experiments were repeated at least twice. Data were first analyzed by one-way ANOVA to determine significant differences between groups; groups were compared using Bonferonni's modification of Student's t-test. A $p < 0.05$ was considered significant.

Phospholipase C Activation

Endothelial cells were seeded in Falcon 6-well culture dishes. When confluent cells were incubated 48hr with myo-[2-³H]inositol (10μCi/ml). The labeled cells were washed three times with PBS then incubated 10 min with PBS containing 10 mM LiCl. The cells were washed once, then incubated with ATP or other agonist as indicated. The reaction was stopped by the addition of 15% trichloroacetic acid and the analysis of the [^{3}H] inositol phosphates carried out as previously described (6,7).

Pertussis Toxin Treatment

Media from confluent RCME cells were removed and the cells washed 3 x with 1 ml of PBS. Cells were incubated four hours in DME containing 0.2% bovine serum albumin, without (control) or with 100ng/ml pertussis toxin.

Permeabilized Cells

RCME cells were permeabilized with 10 μg/ml saponin in a "cytoplasmic" buffer comprised of 20 mM NaCl, 0.5mM MgCl$_2$ H$_2$O, 102 mM KCl, 2.5 mM NaHCO$_3$, 0.96mM NaH$_2$PO$_4$, 1 mM EGTA, 10 mM HEPES and 0.46 mM CaCl$_2$(8).

RESULTS

As reported previously (9), ATP stimulates PG release from RCME cells in a dose-dependent manner. The maximal release of PGs was observed with 500μM ATP (Figure 1). After addition of ATP to the cells, there was a lag of 30-40s before PG release over basal levels could be detected (data not shown). Maximal PG release occurred 4-6 min after adding ATP and levels in the incubation buffer thereafter remained constant for at least 30 min. Therefore, unless otherwise indicated, incubations with ATP were at 500μM for 15 min.

ATP stimulated PLC activation was indicated by the increased production of IP3 and IP2 and the accumulation of IP1 (Figure 2). In the experiments described below, IP1 accumulation at 5 min was used as an index of PLC activation.

Pretreatment of RCME cells with 100 ng/ml pertussis toxin significantly reduced ATP stimulated prostaglandin release but had no effect on [^{3}H] IP1 accumulation (Table 1).

To determine whether PG release could be evoked directly by G-protein activation, we evaluated the effects of GTPγS on PG release in permeabilized cells. As shown in Figure 3, 100 μM GTPγS evoked a significant increase in the release of both PGE$_2$ and 6-keto PGF$_\alpha$. In a parallel series of experiments, GTPγS increased IP1 accumulation (Table 2). Incubation of intact RCME cells with AlF$_4^-$ elicited significant increases in both PGE$_2$ and 6-keto-PGF$_\alpha$ (Figure 4).

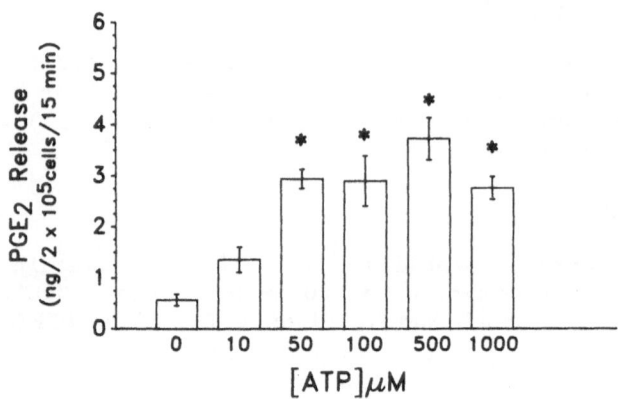

Figure 1. ATP stimulation of prostaglandin release from RCME cells. Cells were incubated with the indicated concentration of ATP for 15 min at 37°C, the incubation buffer collected and prostaglandin content determined by radioimmunoassay. Data are expressed as the mean ± standard error of the mean, n=4. *significantly different from PBS control.

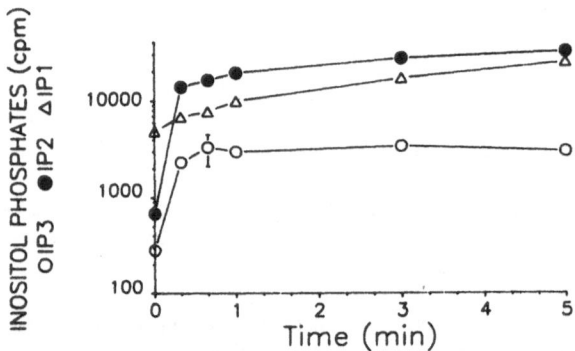

Figure 2. Effect of ATP (500μM) on inositol polyphosphate production in RCME cells. Each point represents the mean ± standard error (n=3). In most cases, the standard error was smaller than the symbol.

Table 1

Effects of Pertussis Toxin on ATP Stimulated
Prostaglandin Release and [^{3}H]IP1 formation

Incubation	PGE$_2$ Release (ng/2 x 10^5 cells/15 min)	[^{3}H]IP1 Formation (cpm at 5 min)
PBS	0.10±0.06	5904±573
Pertussis Toxin*	<0.07	5813±408
ATP (500μM)	1.17±0.29§	30110±408
ATP + Pertussis Toxin	0.51±0.11"	28470±233

Data are expressed as the mean ± standard error of the mean, n=4
(PGE$_2$ release) or n=3 (IP1).
* Cells were pre-incubated with 100ng/ml pertussis toxin for 4 hr
prior to challenge with PBS or ATP.
§ Significantly different from PBS control.
" Significantly different from untreated ATP stimulated cells.

Table 2

Effects of GTPγS on [^{3}H]IP1 formation

Incubation	[^{3}H]IP1 Formation (cpm at 5 min)
Buffer	14090±1306
GTPγS (0.1μM)	32720±1283*

Data are expressed as the mean±standard error of the mean, n=3.
*significantly different from buffer control.

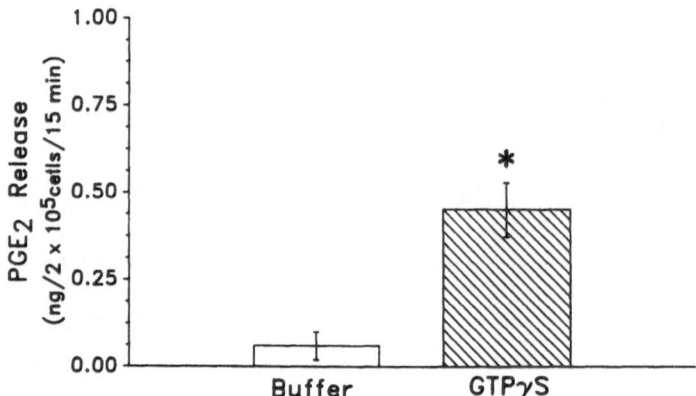

Figure 3. Effects of GTPᴦS(100μM) on PGE₂ release in RCME cells permeabilized with 10 μg/ml saponin. Data are expressed as the mean ± standard error of the mean. (n=4). *significantly different from buffer control.

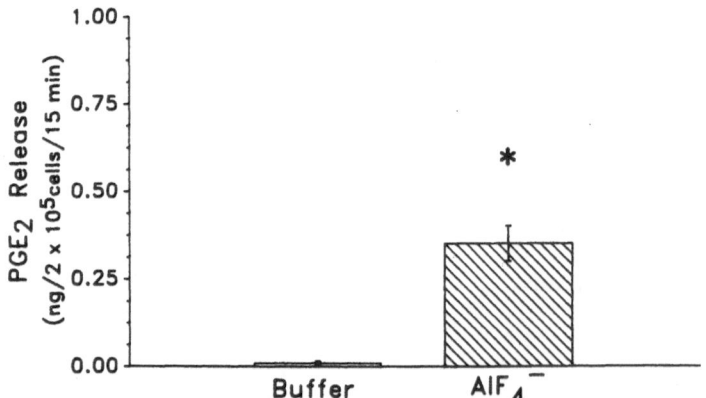

Figure 4. Effects of AlF₄⁻ on PGE₂ production by RCME cells. Confluent cells were incubated with 50μM NaF, 1μM AlCl₃ for 15 min at 37°C. Data are expressed as the mean ± standard error of the mean (n=4). *Significantly different from buffer control.

DISCUSSION

ATP has been reported to stimulate PG release and PLC activity in a number of endothelial cell cultures (10-12) and in all of these studies, ATP actions have been attributed to the P2 purinergic receptors. In RCME cells, ATP stimulation also appears to be via the P2 purinergic receptor since adenosine (1μM-1mM) had no effect on prostaglandin release (data not shown). We believe that PG release serves as a useful index of PLA_2 activation in intact RCME cells since preliminary studies with [^{14}C]arachidonic acid (AA) prelabelled cells indicated that ATP stimulated the release of AA from phosphatidylcholine, phosphatidylethanolamine and phosphatidylinositol. Therefore, although the loss of AA from PI could result from the concerted action of PLC and diglyceride and monoglyceride lipases (13), the release of AA from the other phospholipids strongly suggested that ATP stimulated PLA_2 activation and that the prostaglandins subsequently produced derived, at least in part, from PLA_2 mediated AA release from substrate phospholipids. ATP stimulated phosphatidylinositol turnover in RCME cells as indicated by the increased formation of IP3, IP2 and IP1. However, in contrast to the report of Pirotton et al (12), the levels of all three inositol phosphates remained elevated for 5 min. This discrepancy may be explained by the use of LiCl in our study, which may result in preferential accumulation of 1,3,4 inositol phosphate. IP2 and IP1 accumulation were delayed compared to IP3.

The relationship of PLC activation to PLA_2 activation remains unclear at this time. It is generally thought that PLA_2 activity in endothelial cells is regulated by the intracellular [Ca^{++}](14). PLC activation produces two molecules with second messenger properties, i.e. 1,2 diacylglycerol, which stimulates protein kinase C, and 1,4,5-inositol trisphosphate, which can elicit release of Ca^{++} from intracellular stores (15). Since exogenous IP3 can induce arachidonate formation in permeabilized platelets (15), one possible mechanism for PLA_2 activation is the IP3 stimulated rise in cytosolic Ca^{++} concentration.

The activation of regulatory G-proteins has been associated with many agonist-receptor interactions. The agonist receptor interaction is thought to facilitate the binding of GTP to an α-subunit of the G-protein, resulting in activation. The G-protein then dissociates from the receptor resulting in a decreased receptor affinity and release of the α-subunit. Observations from this study suggest the involvement of G proteins in the activation of both phospholipases. A stable analog of GTP, GTPγS, is a potent and relatively irreversible activator of G-proteins. GTPγS increased IP1 formation and prostaglandin release (16). AlF$_4^-$, a potent activator of G-proteins (16-18) also increased prostaglandin release, and although not evaluated in this study, has been reported to activate PLC in many other cell types (18,19).

ADP ribosylation of certain G-proteins by pertussis toxin irreversibly inactivates them (16). At least two pertussis toxin sensitive G proteins, G_i and G_o, have been identified. Pertussis toxin has been shown to ADP ribosylate a 40-41 kD protein in endothelial cells (20-22) and this ribosylation has been associated with a reduction in LTD_4 stimulated prostaglandin release from bovine pulmonary artery endothelial cells (22). At this time, the identity of the G-proteins ADP-ribosylated by pertussis toxin in endothelial cells remains unknown. In the present study, pertussis toxin treatment reduced ATP stimulated prostaglandin release, but had no effect on ATP stimulated IP1 formation. These observations are similar to reports that pertussis toxin pretreatment does not change the activation of PI turnover in endothelial cells by bradykinin (20,21). Similarly, other investigators have described dissociations of PLC and PLA_2 activation in other receptor-coupled systems (23-26). The lack of effect pertussis toxin on GTP-dependent regulation of PI turnover by ATP suggests that PLC activation by ATP involves a G protein(s) that is not ADP ribosylated by pertussis toxin. In contrast, the ATP-stimulated release of prostaglandins appears to involve a different pertussis toxin sensitive G-protein. Further studies are now needed to identify the putative G-proteins involved in ATP-receptor coupled PLC and PLA_2 activation.

ACKNOWLEDGEMENTS

The authors are grateful to Ms. Carol Perry for her technical assistance. During the period of this study, Dr. Gerritsen was supported by an NIH research Career Development Award.

REFERENCES

1. A. Gilman. G proteins: transducers of receptor-generated signals. Ann. Rev. Biochem. 56: 615-649, 1987.
2. H.R. Bowine and L. Stryer. G proteins: a family of signal transducers. Ann. Rev. Cell. Biol. 2: 391-419, 1986.
3. M.E. Gerritsen and C.D. Cheli. Arachidonic acid and prostaglandin endoperoxide metabolism in isolated rabbit coronary microvessels and isolated cultivated coronary microvessel endothelial cells. J. Clin. Inv. 72: 1658-1671, 1983.
4. M.E. Gerritsen. Eicosanoid production by the coronary microvascular endothelium. Fed. Proc. 46: 47-53, 1987.
5. M.E. Gerritsen, D.M. Nganele and A.M. Rodrigues. Calcium ionophore (A23187) and arachidonic acid stimulated prostaglandin release from microvascular endothelial cells: effects of calcium antagonists and calmodulin inhibitors. J. Pharm. Exp. Ther. 240: 837-846, 1987.

6. I.R. Batty, S.R. Nahorski, and R.F. Irvine. Rapid formation of inositol 1,3,4,5 tetrakisphosphate following muscarinic receptor stimulation of rat cerebral cortical slices. Biochem. J. 232:211-215, 1985

7. C.K. Derian and M.A. Moskowitz. Polyphosphoinositide hydrolysis in endothelial cells and carotid artery segments. J. Biol. Chem. 261: 3831-3837

8. G.M. Burgess, J.S. McKinney, A.Fabiato, B.A. Leslie, and J.W. Putney. Calcium pools in saponin-permeabilized guinea pig hepatocytes. J. Biol. Chem. 258: 15336-15345, 1983.

9. M.E. Gerritsen, C.A.Perry, T. Moatter, E.J. Cragoe, and M.S. Medow. Agonist specific role for Na^+-H^+ antiport in prostaglandin release from microvessel endothelium. Am. J. Physiol. 256: C831-C839., 1989

10. L. Needham, N.J. Cusack, J.D. Pearson and J.L. Gordon. characteristics of the P2 purinoceptor that mediates prostacyclin production by pig aortic endothelial cells. Eur. J. Pharmacol. 134: 199-209, 1987.

11. T.A. Brock, P.A. Dennis, K.K. Griendling, T.S. Diehl, and P.F. Davies. GTPγS loading of endothelial cells stimulates phospholipase C and uncouples ATP receptors. Am. J. Physiol 255: C667-C673, 1988.

12. S. Pirotton, E. Raspe, D. Demolle, C. Erneux, and J.M. Boeynaems. Involvement of inositol 1,4,5-triphosphate and calcium in the action of adenine nucleotides on aortic endothelial cells. J. Biol. Chem. 262: 17461-17466, 1987.

13. S.L. Hong and D. Deykin. Activation of phospholipases A_2 and C in pig aortic endothelial cells synthesizing prostacylin. J. Biol. Chem. 257: 7151-7154, 1982.

14. E.A. Jaffe, E.A., Grulich, J., B.B. Weksler G. Hampel, and K. Watanabe. Correlation between thrombin induced prostacyclin production and inositol trisphophate and cytosolic free calcium levels in cultured endothelial cells. J. Biol. Chem. 262: 8557-8565, 1987.

15. M.J. Berridge and R.F. Irvine. Inositol trisphosphate, a novel second messenger in cellular signal transduction. Nature 312: 315-321, 1984.

16. M. Freissmuth, P.J. Casey, and A.G. Gilman. G proteins control diverse pathways of transmembrane signalling. Faseb. J. 3: 2125-2139, 1989.

17. P. Sternweis and A. Gilman. Aluminum: a requirement for the activation of the regulatory component of adenylate cyclase by fluoride. Proc. Natl. Acad. Sci. USA 79: 4888-4891, 1982.

18. P. Blackmore, S. Bocckino, L. Waynick and J. Exton. Role of a guanine nucleotide-binding regulatory protein in the hydrolysis of hepatocyte phosphatidylinositol 4,5-bisphosphate by calcium mobilizing hormones and the control of cell calcium. Studies using aluminum fluoride. J. Biol. Chem. 260: 14477-14483, 1985.

19. I. Fuse and H.H. Tai. Stimulations of arachidonate release and inositol 1,4,5-triphosphate formation are mediated by distinct G-

proteins in human platelets. Biochem. Biophys. Res. Comm. 146: 659-665, 1987.
20. T.A.Voyno-Yasenetskaya, V.A. Tkachuk, V.A., E.G. Cheknyova, M.P. Panchenko, G.Y. Grigorian, R.J. Vavrek, J.M. Stewart, and U.S. Ryan. Guanine nucleotide-dependent pertussis-toxin insensitive regulation of phosphoinositide turnover by bradykinin in bovine pulmonary artery endothelial cells. Faseb J. 3: 44-51, 1989.
21. T.L. Lambert R.S. Kent and A.R. Whorton. Bradykinin stimulation of inositol polyphosphate production in porcine aortic endothelial cells. J. Biol. Chem. 15288, 15293, 1986.
22. M.A. Clark, T.M. Conway, C.F. Bennett, S.T. Crooke, and J.M. Stadel Islet activating protein inhibits leukotriene D_4 and C_4 but not bradykinin or calcium ionophore-induced prostacyclin synthesis in bovine endothelial cells. Proc. Natl. Acad. Sci. 83: 7320-7324, 1986.
23. L.F. Brass, C.C. Schaller and E.J. Belmonte.Inositol 1,4,5-triphosphate induced granule secretion in platelets. Evidence that the activation of phospholipase C mediated by platelet thromboxane receptors involves a guanine nucleotide distinct from that of thrombin. J. Clin Invest. 79: 1269-1265, 1986.
24. S. Slivka, and P.A. Insel. Alpha 1-adrenergic receptor mediated phosphoinositide hydrolysis and prostaglandin E_2 formation in Madin-Darby canine kidney cells. Possible parallel activation of phospholipase C and phospholipase A_2. J. Biol. Chem. 262: 4200-4207, 1987.
25. R.M. Burch and J. Axelrod.Dissociation of bradykinin induced prostaglandin formation from phosphatidylinositol turnover in Swiss 3T3 fibroblasts: evidence for G protein regulation of phospholipase A_2. Proc. Natl. Acad. Sci. 83: 6374-6378, 1987.
26. M.F. Crouch and E.G. Lapetina. No direct correlation between Ca^{++} mobilization and dissociation of G_i during platelet phospholipase activation. Biochem. Biophys. Res. Commun. 153: 21-30, 1988.

THE ROLE OF PHOSPHOLIPASE A_2 ACTIVATING PROTEIN (PLAP) IN REGULATING PROSTANOID PRODUCTION IN SMOOTH MUSCLE AND ENDOTHELIAL CELLS FOLLOWING LEUKOTRIENE D_4 TREATMENT

Mike A. Clark*, John S. Bomalaski°, Theresa M. Conway, Mike Cook, Janice Dispoto, Seymore Mong, Robert G.L. Shorn[+], Jeff Stadell, Lynne Webb*, and Stanley T. Crooke**

Smith Kline and French, Philadelphia, Pennsylvania
*Washington University School of Medicine, St. Louis, Missouri, Current address: Schering-Plough Research, Bloomfield, New Jersey
°Veterans Administration Medical Center, Medical College of Pennsylvania, Philadelphia, Pennsylvania
[+]Current address: AT Biochem, Malvern, Pennsylvania
**Current address: ISIS, San Diego, California

Leukotrienes are a family of compounds originally termed the slow-reacting substances of anaphylaxis and are recognized as important mediators of anaphylaxis (1). In particular, leukotriene C_4 (LTC_4) and leukotriene D_4 (LTD_4) are potent spasmogens in a variety of smooth muscle tissues including trachea, lung, ileum and the vasculature (2.4). The mechanism by which contraction is induced by the leukotrienes (LTs) is not yet known. However, in some experimental systems, but not all, a cyclooxygenase product of arachidonic acid metabolism, possibly thromboxane B_2 (TxB_2), may be involved in mediating leukotriene-induced effects (3,4). This is based on the observation that inhibitors of cyclooxygenase, such as indomethacin and meclofenamic acid, can block leukotriene-induced contraction of the lung parenchyma, vasculature and the ileum (2.4). In other tissues, most notably guinea pig trachea, the cyclooxygenase products appear to be relatively unimportant mediators of leukotriene effects (5). Furthermore, in the tissues in which cyclooxygenase products appear to be important, thromboxane synthesis may be crucial for the leukotriene responses (4). Although TxB_2 is usually assumed to be of platelet origin, the cellular source of LT-induced thromboxane synthesis is not known. This is largely because a homogeneous smooth muscle system for studying the effects of LTs has not yet been described. Previous investigators have employed in vivo and organ-bath systems, and therefore, the specific cell types that respond to LTs and produce TxB_2 have been difficult to identify.

Phospholipase A2
Edited by P.Y.-K. Wong and E. A. Dennis
Plenum Press, New York, 1990

125

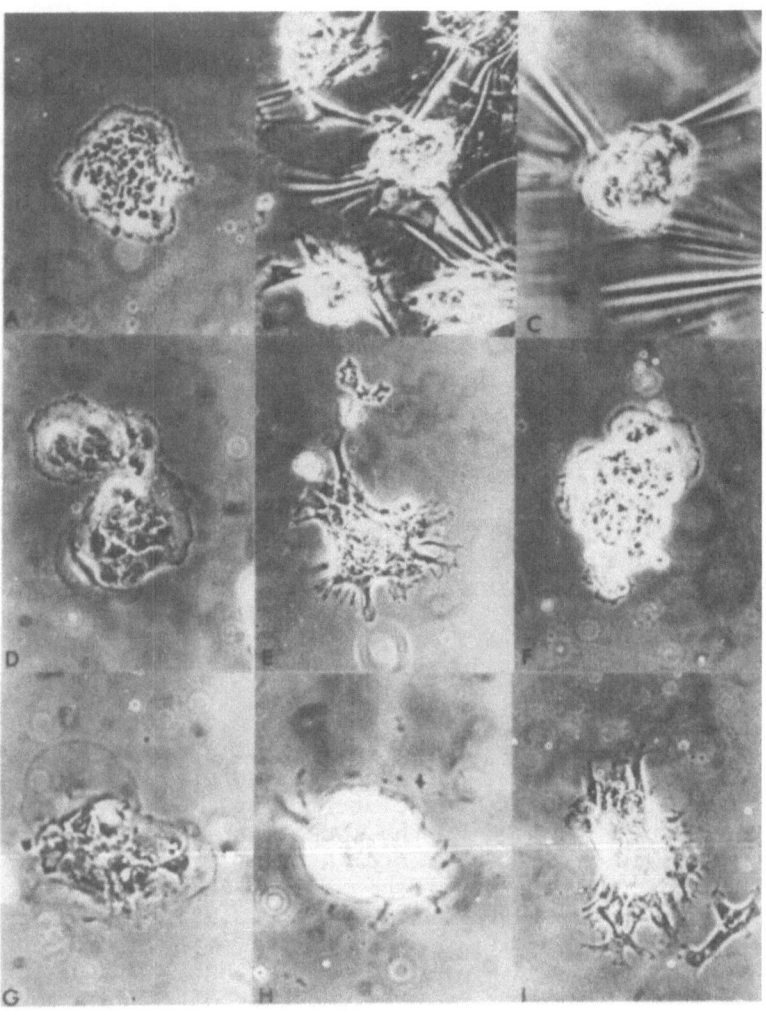

Figure 1. LT-induced smooth muscle cell contraction. The latex substrate was used to demonstrate cell contraction in response to leukotriene treatment. Figure 1A illustrates control culture. Figure 1B, approximately 3–5 min after the addition of 10 nM of LTC_4 many wrinkles were observed. Figure 1C, approximately 3–5 min after the addition of 1 μM LTD_4 contractions were also observed. Figure 1D, 5R,6S–LTD_4 (10 μM) and Fig. 1E, LTE_4 (10 μM) failed to induce contraction. Pretreatment of the cells with the cyclooxygenase inhibitor meclofenasmate (10 μM) for 5 min inhibited contraction in response to LTC_4 (1 μM; fig. 1F) and LTD_4 (10 μM; fig. 1G). In addition, the thromboxane synthetase inhibitor benzylimidazole (10 μg/ml for 5 min) also blocked LTC_4 (1 μM) (fig. 1H) and LTD_4 (10 μM) (fig. 1) induced contraction. Bar=10 μm in each figure.

We have chosen to examine the binding of LTs and to study their effects on cells in tissue culture for several reasons. First, a homogeneous cell population allows the identification of specific cell types involved in the LT- mediated response. Second, if specific receptor sites and biological effects are identified, they can be shown to have occurred in the same cell type.   Third, in vitro systems allow easier manipulation.   Fourth, such systems are conducive to more precise analysis of potential second messengers and other intracellular effects.

We observed that $BC3H_1$ smooth muscle cells contract (e.g. exert a mechanical force as detected by the wrinkles in a rubber substrate) following treatment with LTD (Figure 1) (5). Concurrent with the observed contraction in response to LTC_4 and LTD_4 was an increase in the amount of TxB_2 in the supernatant of treated cells. Using radioimmunoassays, it was possible to demonstrate a significant increase in TxB_2 secretion over control levels using concentrations as low as 1 μM and 10 nM of LTD_4 or LTC_4, respectively (Figure 2). Higher doses of leukotriene resulted in dose-dependent increases in the amount of TxB_2 detected in the supernatant.   The 5R,6S stereoisomer of LTD_4 was

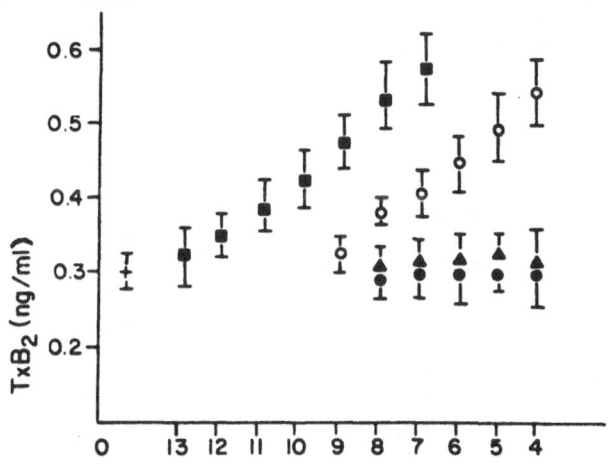

CONCENTRATION (-log M)

Figure   2. The effects of the various leukotrienes on thromboxane synthesis. Supernatants obtained from the $BC3H_1$ after 20 min of incubation with LTs at 37°C were assayed by RIA. (+) Control (■) LTC_4, (O) LTD_4, (▲) LTE_4, (●) 5R,6S-LTD_4. Data shown are the averages of 3 experiments (mean ± S.D.). Both LTC_4 and LTD_4 produced dose dependent increases in TXB_2 synthesis after 20 min of incubation.

at least 100-fold less efficient than 5S,6R–LTD$_4$ in inducing TxB$_2$ secretion, demonstrating the stereospecific response to these cells to LTD$_4$. Meclofenamic acid blocked TxB$_2$ synthesis and also blocked the wrinkling of the latex substratum in response to LTC$_4$ and LTD$_4$ (Figure 1). These results suggest that the mechanism of LTC$_4$– and LTD$_4$–induced contraction in BC3H$_1$ cells involves cyclooxygenase products. This hypothesis was supported by the observation that the thromboxane synthetase inhibitor, benzylimidizol, also blocked LTC$_4$ and LTD$_4$–induced contraction. Thromboxane A$_2$ (TxA$_2$), a cyclooxygenase product, is noted for its ability to contract smooth muscle.

The significant increase in TxB$_2$ following LTD$_4$ treatment described here suggests that the cyclooxygenase product involved in the LTD$_4$–induced contraction of these cells might be TxA$_2$. This hypothesis is further supported by the data showing that these cells contracted in response to TXA$_2$. It has been demonstrated in both organ bath as well as in vivo systems that a major portion of the LTC$_4$– and LTD$_4$–induced contraction of lung, uterus, vascular and ileal smooth muscle can be blocked by cyclooxygenase inhibitors (3,4). In contrast, LTC$_4$ and LTD$_4$ induced contraction of tracheal smooth muscle even in the presence of cyclooxygenase inhibitors (5). Therefore, it would appear as if the BC3H$_1$ cells are more similar to the smooth muscle cells of the vasculature, lung and ileum than to the smooth muscle cells in the trachea in their responses to LT. Thus these cells appear to be of use as a model system for studying the cyclooxygenase-mediated effect of leukotriene treatment.

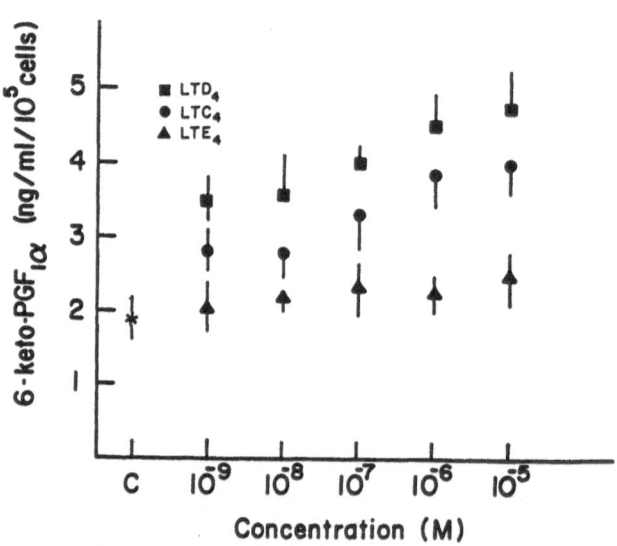

Figure 3. Prostacyclin secretion into the medium of the cells was measured by RIA for 6–keto–PGF$_{i3}$ after 10 min of treatment.

We have also studied the effects of LTD_4 on the bovine endothelial cell line CPAE (6). This cell line produces eicosanoids in response to not only LTD_4 but also bradykinin (Figure 3) and thus we can ask if the signal transduction pathways for LTD_4 and bradykinin are the same in this cell line. Peptidoleukotrienes stimulate the synthesis of prostanoids in both smooth muscle and endothelial cells. Since the rate-limiting step in the synthesis of eicosanoids appears to be the release of arachidonic acid from phospholipids (8), our initial studies focused on the molecular mechanism of peptidoleukotriene-induced stimulation of eicosanoid synthesis.

Our early studies demonstrated that stimulation of the $BC3H_1$ and CPAE cell lines with LTD_4 resulted in a substantial augmentation in prostanoid production as determined by radioimmunoassays in which LTD_4 could have augmented prostanoid production in these cells. The first possibility was that LTD_4 was specifically directing released arachidonic acid into a particular prostanoid in preference to other metabolic pathways (e.g. lipooxygenase or arachidonyl CoA synthetase).

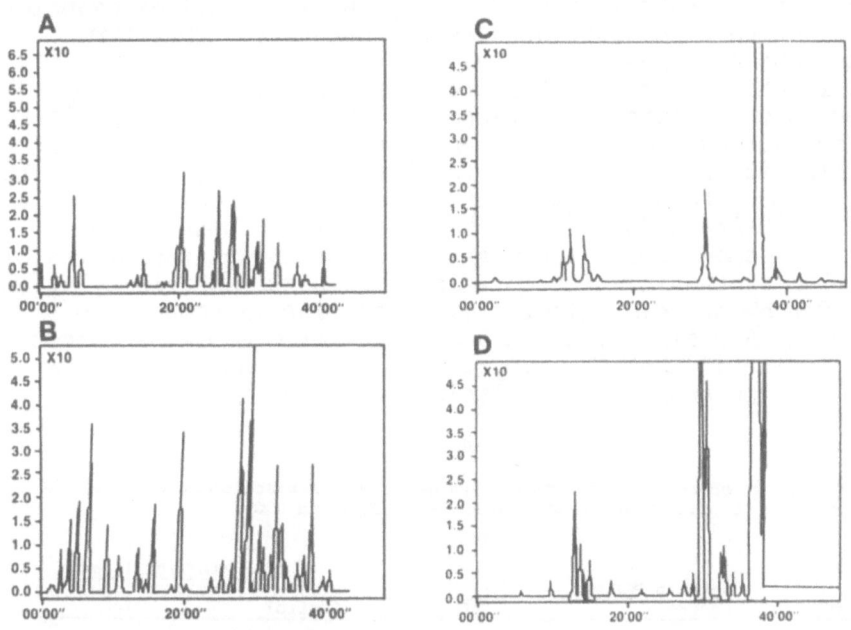

Figure 4. HPLC Analysis of [^{3}H]-Arachidonic Acid Metabolites Produced by the cells. Cells (5×10^6) were incubated with [^{3}H]-arachidonic acid (10uCi/ml) for 14 hr, washed and subsequently incubated in saline (A) or saline containing LTD_4 (1 uM) for 10 min (B). The supernatant was extracted and analyzed by reverse phase HPLC using an on-line radioactivity detector. The chromatograms are representative examples obtained from 2 experiments.

The major peaks were identified by comparisons of their retention times with authentic standards and their integrated areas are shown below.

Alternatively, LTD_4 was increasing the rate of release of arachidonic acid which could then be converted into all prostanoids. To distinguish between these possibilities, cell phospholipids were prelabeled by incubation for 24 hr in the presence of ^{3}H-arachidonic acid (10 uCi/ml). The cells were rinsed three times with saline and subsequently incubated for an additional 10 min in either saline or saline containing LTD_4 (1 um). The supernatants were collected, acidified with phosphoric acid (0.1% final concentration) and the radioactive eicosanoids were extracted with ethylacetate. Greater than 95% of the total radioactivity released by the cells was recovered by this procedure. Radiolabeled eicosanoids were concentrated and separated into individual molecular constituents using reverse phase HPLC (Figure 4) (7). Treatment with LTD_4 resulted in an increase in the release of not only prostanoids but also increased the release of a variety of other eicosanoids concomitant with the accumulation of free arachidonic acid (Figure 4) (9). Taken together, these results indicate that LTD_4 did not selectively shunt arachidonic acid into the production of a particular prostanoid but rather increased the synthesis of all eicosanoids, a result which is consistent with the hypothesis that LTD_4 increased phospholipase activity.

To identify whether a G protein was coupled to the LTD_4 receptor which mediated this response, the pertussis toxin, IAP (islet-activating protein), was employed (8). Pretreatment of the CPAE cells with IAP (100 ng/ml) for 1 hr completely inhibited the ability of LTD_4 to stimulate prostacyclin production and the release of ^{3}H-arachidonic acid (Table 1) (10). Dose-response experiments demonstrated a concentration-dependent inhibition of prostacylin synthesis which correlated with the ADP ribosylation of a 41,000 MW protein (Figure 5). Since IAP is thought to specifically ADP ribosylate only a subset of G proteins these results indicated that the LTD_4 receptor was coupled though a pertussis toxin-sensitive G protein, possibly Gi.

TABLE 1

The effect of IAP on prostacyclin synthesis and ^{3}H-arachidonic acid release in response to bradykinin, ionophore A23187, LTC_4, and LTD_4.

Treatment	6-keto PGF_1 Control	IAP	^{3}H arachidonic acid release (cpm) Control	IAP
None	2.4 + 0.6	2.4 + 0.2	500	600
LTD_4 (luM)	4.6 + 0.2	2.5 + 0.4*	2000	550
LTC_4 (luM)	4.1 + 0.4	1.9 + 0.3*	2300	500
Bradykinin (2uM)	4.3 + 0.2	3.8 + 0.4+	2200	2300
Ionophore A23187 (luM)	4.2 + 0.1	3.9 + 0.2+	2500	2300

Cells were pretreated with IAP (100ng/ml) for 1 hr prior to being treated with bradykinin (2uM), inophore (luM), LTD_4 (luM), or LTC_4 (luM). After an additional 10-min incubation, the supernatants were assayed for 6-keto-PGF_1 (mg/ml) by radioimmunoassay. The data shown was obtained from three experiments (mean + SD). *P < 0.001. +Not significant, P > 0.1. The right hand column denotes release of [^{3}H] arachidonic acid metabolites from CPAE cells pretreated in the absence or presence of IAP. CPAE cells were cultured overnight in the presence of [^{3}H] arachidonic acid at 10uCi/ml. The incubation was continued for an additional 1 hr in the absence or presence of IAP

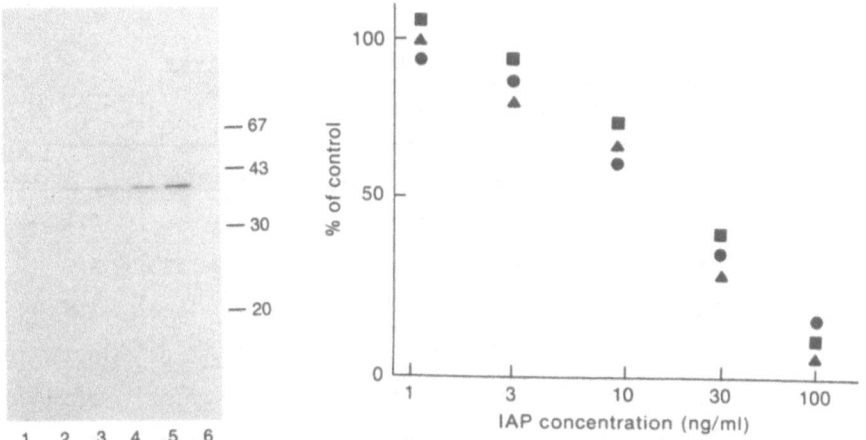

Figure 5. [^{32}P]ADP-Ribosylation Catalyzed by Islet Activating Protein in CPAE cells. CPAE cells were preincubated with various concentrations of IAP (100, 20, 10, 3, 0, and Ong: lanes 1-6, respectively) for 1 hr at 37^{0}C. To demonstrate ribosylation in the intact cell, membranes were prepared from the treated cells and subsequently incubated with 5uM [^{32}P]NAD in the presence (lanes 1-5) and absence (lane 6) of additional IAP (5ug/ml). The membranes were dissolved in NaDodSO$_4$ sample buffer and electrophoresis was carried out on a 10% polyacrylamide gel. The gel was dried and bands were visualized by autoradiography. The positions of molecular weight standards (in kDa) are indicated at the right. As can be seen in lane 1, essentially all of the 42K protein was ribosylated in the cell since essentially no additional ribosylation occurred in isolated membranes.

Subsequently, to examine the effect of IAP on prostacyclin production in response to leukotriene treatment, radioimmunoassays were preformed and the results were correlated with the ribosylation of the 42k protein. Gels were sliced and the radioactivity of the 42-kDa protein was determined by scintillation spectrometry (●). These data are plotted with data obtained from radioimmunoassays for 6-keto-PGF$_1$ produced by cells treated with luM LTD$_4$ (■) and luM LTC$_4$ (▲). The radioimmunoassays were performed using cells that had been pretreated with various concentrations of IAP for 1 hr, then luM LTD$_4$ was added to the cell cultures, and supernatants were assayed 10 min later. The radioimmunoassay data were obtained from three separate experiments, and the [^{32}P]ADP-ribosylation data were obtained from two experiments.

        Pretreatment of those cells with either cycloheximide or actinomycin D resulted in complete ablation of the augmentation of prostacyclin or TxB$_2$ accumulation, although no effects on the conversion of exogenously added arachidonic acid to prostacyclin were observed (Table 2) (5,6). These results confirmed earlier reports demonstrating that the availability of free arachidonic acid was the rate-limiting step in eicosanoid formation (9). Furthermore, these studies suggested that protein synthesis was necessary for the release of arachidonic acid from endogenous phospholipid storage pools but that protein synthesis was not necessary for the oxidation of arachidinic acid into prostanoids. This suggested to us that whole cell sonicates made from cells which were pretreated with LTD$_4$ should have enhanced phospholipase activity.

TABLE 2

EFFECTS OF PROTEIN AND RNA SYNTHESIS INHIBITORS
ON PROSTACYCLIN SYNTHESIS IN CPAE CELLS AND
TXBZ SYNTHESIS IN BC3H$_1$ CELLS

Culture Conditions	Cell Line	None	Cycloheximide	INHIBITOR Actinomycin D
Control	CPAE	2.4 ± .6	1.4 ± .5	2.1 ± .3
+LTD$_4$ (1 uM)	CPAE	4.6 ± .2	1.9 ± .3**	1.9 ± .2**
+Arachidonic Acid	CPAE	3.7 ± .2	3.2 ± .3	3.5 ± .2
Control	BC3H$_1$	0.3 ± 0.03	0.27 ± 0.02	0.26 ± 0.03
+LTDy	BC3H$_1$	0.43 ± 0.04	0.19 ± 0.02	0.28 ± 0.07
Arachidonic Acid	BC3H$_1$	0.41 ± 0.04	0.39 ± 0.04	0.42 ± 0.04

Values expressed are ng/ml/10^5 cells/ 10 min of 6-keto-PGF$_{1\alpha}$. Values significantly differ (**$p < 0.05$) from the control samples.

Cells were pretreated with either saline, or saline containing cycloheximide (100 ug/ml) or saline containing actinomycin D (10ug/ml) for 30 min at 37°C. Next the cells were rinsed with fresh saline (containing the appropriate inhi
bitor) and treated with LTD$_4$ or arachidonic acid for 10 min at 37°C. The supernatants were then collected and assayed for prostacyclin on TXBZ by radioimmunoassay. The concentrations of cycloheximide and actinomycin D used were sufficient to inhibit by 95% the incorporation of [^{3}H]-Leu and [^{3}H]-U respectively yet did not affect cell viability as determined by trypan blue dye exclusion during the course of these experiments.

To determine which phospholipids contained radiolabeled arachidonic acid, lipids were isolated by chloroform and methanol extraction and subsequently separated by two dimensional thin-layer chromatography. The major phospholipid pools which contained radiolabeled arachidonic acid were phosphatidylcholine, phosphatidylethanolamine and phosphatidylinositol. Because LTD$_4$ resulted in the release of only 2% of the total incorporated arachidonic acid, we could not conclusively demonstrate which lipid pool was the source of the released arachidonic acid.

To identify the biochemical mechanisms responsible for the release of arachidonic acid we assayed phospholipase A$_2$ and phospholipase C enzyme activities in control and LTD$_4$ treated cells using these three separate phospholipid classes as substrate (Table 3a and 3b) (10). The strategy underlying these experiments was to compare phospholipase activities in cell-free preparations from control and treated cells. Briefly each lipid substrate was radiolabeled in the sn-2 position with ^{14}C-arachidonic acid and in the polor head group with ^{3}H. In all of the experiments described in this section the substrates used were 1-palmitoyl-2-[^{14}C]-arachidonyl-sn-glycero-3-[choline-methyl-^{3}H]-phosphocholine, 1-palmitoyl- 2-[^{14}C]-arachidonyl-sn-glycero-3-phospho-[^{3}H]-ethanolamine and 1-stearoyl-2-[^{14}C]-arachidonyl-sn-glycero-3-

phospho-myoinositol-[^{3}H]. Phospholipase A$_2$ activity was
quantitated by the formation of [^{3}H]-lysophospholipid (2,000–
20,000 cpm) and phospholipase C activity was quantitated by the
formation of [^{14}C] diglyceride (500 – 2,000 cpm). All reactions
were performed under conditions in which the reaction product
formation was linear with respect to time and protein. In
addition we also confirmed phospholipase A$_2$ activity values by the
formation of [^{14}C] free fatty acid (arachidonic acid) and
confirmed measured phospholipase C activity by the formation of
[^{3}H] phosphocholine, [^{3}H] phosphoethanolamine or [^{3}H]-inositol.
These later data were used to confirm the previous values as these
metabolites could have been produced by other sequential
mechanisms (i.e. [^{14}C] free fatty acid formation could have
resulted from sequential phospholipase C and mono- and diglyceride
lipase activities). We found that LTD$_4$ increased phospholipase A$_2$
activity but had no effect on phospholipase C activity.

TABLE 3a

Effects of leukotriene D$_4$ on phospholipase A$_2$ and phospholipase C
activities in smooth muscle cells (BC3H$_1$)

Cells were treated for 3 min with leukotriene D$_4$ (1 μM) or
(5R,6S)-leukotriene D$_4$ (1 μM) and then sonicated, and the phospholipase
activites were measured. The assays for phospholipase A$_2$ activity and for
phospholipase C activity were done in cell-free preparations as described under
"Materials and Methods." In some of the experiments, cycloheximide (100 μg/ml)
or actinomycin D (10 μg/ml) was added to the cells 10 min prior to the addition
of leukotriene D$_4$. Values expressed are in picomoles of product produced per
min/mg of protein. The data represent the results from three experiments
assayed in triplicat (mean $\pm$ S.D.). BG, activity not detected; ND, not
determined; PLA$_2$, phospholipase A$_2$; PLC, phospholipase C; PC,
phosphatidylcholine; PI, phosphatidylinositol; PE, phosphatidylethanolamine.

Activity	Substrate	Inhibitor	Control	+LTD$_4$ (1 μM)	+(5R,6S)-LTD$_4$ (1 μM)
PLA$_2$	PC	None	8.8 $\pm$ 1.6	17.0 $\pm$ 4.0	7.6 $\pm$ 3.4
		Cycloheximide	6.2 $\pm$ 1.4	6.2 $\pm$ 0.6	
		Actinomycin D	9.0 $\pm$ 2.8	8.0 $\pm$ 0.7	
	PI	None	BG	BG	BG
		Cycloheximide	BG	BG	
		Actinomycin D	BG	BG	
	PE	None	0.25 $\pm$ 0.13	0.23 $\pm$ 0.08	0.15 $\pm$ 0.05
		Cycloheximide	0.35 $\pm$ 0.04	0.31 $\pm$ 0.13	
		Actinomycin D	0.24 $\pm$ 0.02	0.26 $\pm$ 0.07	
PLC	PC	None	16.5 $\pm$ 5.0	19.6 $\pm$ 6.0	17.6 $\pm$ 3.0
		Cycloheximide	15.0 $\pm$ 5.0	13.0 $\pm$ 5.0	
		Actinomycin D	12.0 $\pm$ 2.0	9.5 $\pm$ 2.0	
	PI	None	21.0 $\pm$ 14.0	19.0 $\pm$ 9.0	21.0 $\pm$ 3.0
		Cycloheximide	22.0 $\pm$ 8.0	33.0 $\pm$ 5.0	
		Actinomycin D	28.0 $\pm$ 8.0	29.0 $\pm$ 7.0	
	PE	None	BG	BG	ND
		Cycloheximide	BG	BG	
		Actinomycin D	BG	BG	

TABLE 3b

Effects of leukotriene D$_4$ on phospholipase A$_2$ and phospholipase C activities in endothelial cells (CPAE)

Cells were treated for 3 min with leukotriene D$_4$ (1 μM) or (5R,6S)-leukotriene D$_4$ (1 μM) and then sonicated, and the phospholipase activities were measured. The assays for phospholipase A$_2$ activity and for phospholipase C activity were done in cell-free preparations as described under "Materials and Methods." In some of the experiments, cycloheximide (100 μg/ml) or actinomycin D (10 μg/ml) was added to the cells 10 min prior to the addition of leukotriene D$_4$. Values expressed are in picomoles of product produced per min/mg of protien. The data represent the results from three experiments assayed in triplicate (mean $\pm$ S.D.). See the legend to Table I for definitions of abbreviations.

Activity	Substrate	Inhibitor	Control	+LTD$_4$ (1 μM)	+(5R,6S)-LTD$_4$ (1 μM)
PLA$_2$	PC	None	7.4 $\pm$ 3.5	32.0 $\pm$ 3.1	7.6 $\pm$ 3.4
		Cycloheximide	8.1 $\pm$ 5.1	5.2 $\pm$ 3.7	
		Actinomycin D	6.4 $\pm$ 1.4	6.3 $\pm$ 4.9	
	PI	None	BG	BG	BG
		Cycloheximide	BG	BG	
		Actinomycin D	BG	BG	
	PE	None	2.1 $\pm$ 1.5	1.7 $\pm$ 2.0	2.0 $\pm$ 1.0
		Cycloheximide	4.7 $\pm$ 3.0	4.0 $\pm$ 2.0	
		Actinomycin D	1.9 $\pm$ 0.5	3.6 $\pm$ 1.2	
PLC	PC	None	6.2 $\pm$ 8.0	9.5 $\pm$ 3.0	6.4 $\pm$ 4.0
		Cycloheximide	7.1 $\pm$ 3.5	6.1 $\pm$ 2.9	
		Actinomycin D	7.6 $\pm$ 2.5	7.2 $\pm$ 2.2	
	PI	None	38.7 $\pm$ 5.2	48.0 $\pm$ 2.3	31.0 $\pm$ 3.0
		Cycloheximide	37.0 $\pm$ 3.0	45.0 $\pm$ 13.0	
		Actinomycin D	37.0 $\pm$ 1.4	42.0 $\pm$ 5.0	
	PE	None	BG	BG	ND
		Cycloheximide	BG	BG	
		Actinomycin D	BG	BG	

Figure 6. Effect of Leukotriene D$_4$ on the Induction of Phospholipase A$_2$ Activity and Prostanoid Synthesis. Cells were treated with leukotriene D$_4$ (1uM) for the indicated times and assayed in triplicate in a cell-free preparation for phospholipase A$_2$ (PLA$_2$) activity using 1-palmitoyl-2-[^{14}C]-arachidonyl-sn-glycero-1-[choline-methyl-H]-phosphocholine as substrate. Prostanoid synthesis was quantitated at the indicated times by radioimmunoassay. The values in these experiments were determined as percent increases above control values for phospholipase A$_2$ activity (o) or 6-keto PFG production (X). The data shown represent the (mean $\pm$ S.D.) from three experiments assayed in triplicate. The control values for PLA$_2$ and prostacyclin production are 7.4pmol of lysopholipid produced per min per mg of cell protein and 2.4ng produced per ml per 10^5 cells per 10 min respectively.

Next, we directly measured phospholipase A_2 activity in BC3H$_1$ and CPAE cell sonicates as a function of time after leukotriene D$_4$ treatment. A three-fold increase in phospholipase A_2 activity was observed within two minutes of LTD$_4$ stimulation (Figure 6). Phospholipase A_2 activation directly preceded the formation of prostanoids in these cells. Thus, the three-fold increase in measurable phospholipase activity <u>in vitro</u> was observed just prior to the accumulation of oxygenated eicosanoid metabolites in the cell supernatants.

To determine the concordance between the increase in measurable phospholipase A_2 activity and the accumulation of 6-keto-PGF$_1$, the concentration-dependence of these phenomena and their specificities were examined. Incubation of cells with the biologically active isomer of leukotriene D$_4$ (5S, 6R) resulted in a statistically significant increase in both measurable phospholipase A_2 activity as well as prostacyclin accumulation. This phenomena was dose-dependent and increases in both of these responses were observed at 100 nM and 1 uM leukotriene D$_4$ (Figure 7). This effect was saturable since increasing LTD$_4$ concentration above 1 uM had no additional effects on either of these responses. Treatment of the BC3H$_1$ and CPAE cell lines with (5R, 6S) LTD$_4$, the biologically <u>inactive</u> isomer, failed to increase either measurable phospholipase A_2 activity or eicosanoid accumulation even when present at 1 uM concentration. This result suggested that these responses were receptor-mediated. This was confirmed using specific receptor antagonists.

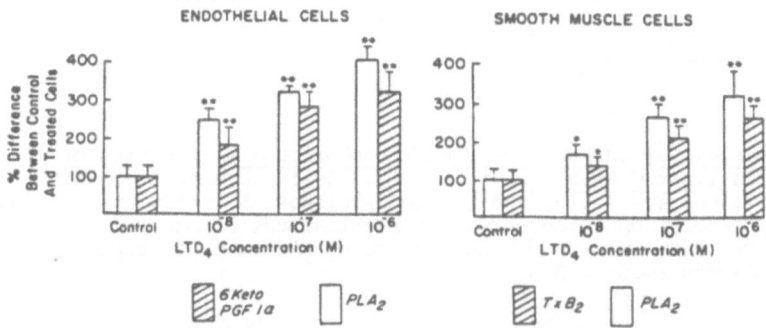

Figure 7. <u>Effect of Increasing Concentrations of Leukotriene D$_4$ on Phospholipase A2 Activity and on 6-Keto-PGF$_1$ Production.</u> Cells were treated with leukotriene D$_4$ for 3 min at the indicated concentrations and assayed for phospholipase activity in a cell-free preparation with phosphatidylcholine (1-palmitoyl-2-[^{14}C]-arachidonyl-sn-gylcero-3-phosphocholine) as substrate. The synthesis of 6-keto-PGF$_{1a}$ and TXBZ was determined by radioimmunoassays (New England Nuclear) following a 10-min treatment of cell lines with leukotreine D$_4$. Values in these experiments were expressed as percent increase above controls. Data are presented as a percent of control where the control levels of 6-keto-PGF$_{1a}$ were 2.0ng produced in 10 min by 10^5 endothelial cells and the control levels of phospholipase A_2 activity were 7.4pmol of phosphatidylcholine hydrolyzed per min/mg of protein. The data shown represent results obtained from three experiments assayed in triplicate (mean $\pm$ S.D.). **,p < 0.05.

To determine whether the observed increases in measurable
phospholipase A_2 activity were related to alterations in the
maximal velocity or substrate binding, Lineweaver–Burk analysis of
the kinetic data was employed. Treatment of endothelial and
smooth muscle cells with leukotriene D_4 for three minutes resulted
in an increase in the maximum velocity of fatty acid release from
phosphatidylcholine substrate while no alteration in the binding
affinity was observed (Figure 8). These results demonstrated that
phospholipase A_2 activity is specifically augmented by treatment
of these cells with leukotriene D_4 in a fashion which is
concentration–dependent, saturable, stereospecific and is
temporally related to prostacyclin synthesis.

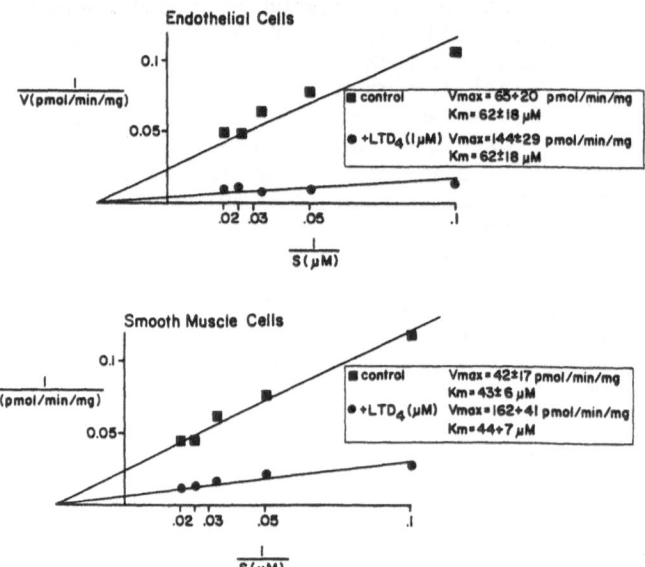

Figure 8. Lineweaver Burk Analysis of Phospholipase A_2 Activity of
Endothelial Cells Treated with Leukotriene D_4. Cells were treated with
leukotriene D_4 (1 uM) for 3 min, and the phospholipase A_2 activity was analyzed
in a cell-free preparation using varying concentrations of phosphatidylcholine
as substrate. The data shown represent the results from three experiments
assayed in triplicate (mean $\pm$ S.D.).

Taken together, these data suggest that a protein is made
soon after leukotriene D_4 treatment which enhances the maximum
velocity of phospholipase A_2. This protein was originally
postulated to be either a phospholipase or a protein which can
accelerate endogenous phospholipase activity (either by activation
or by ablation of inhibition). Since phospholipase activating
proteins are abundant constituents of insect venoms we speculated
that a similar protein might be synthesized after agonist
stimulation in the $BC3H_1$ and CPAE cell line. To directly explore
this possibility, we generated antibodies to the bee venom
constituent mellitin by cross-linking mellitin with gluteraldehyde
to form insoluble complexes which were subsequently injected
intradermally into New Zealand white rabbits (12). The resulting
antibodies were covalently bound to a HPLC immunoaffinity column

and, subsequently, cytosol was percolated through the column. The
column was extensively washed and specifically bound protein was
eluted by application of pH 3.1 Na acetate buffer. Column eluents
were assayed for phospholipase A_2 stimulatory activity. A single
peak of stimulatory activity was found (Figure 9) (11). To
further purify this polypeptide, gel filtration chromatography was
utilized. The fractions containing stimulatory activity were
loaded onto a 1 x 24 cm HPLC gel filtration column and eluents
were again assayed for phospholipase stimulatory activity. The
majority of stimulatory activity eluted as a single peak with an
apparent molecular mass of 43 kD (Figure 10). We have included
0.05% Tween 20 in the purification of PLAP for the following
reasons: 1) this amount of detergent increases (by about 20 –
40%) the yield of PLAP; and 2) use of a detergent during affinity
purification reduces nonspecific interactions of proteins with the
affinity column thus reducing the amount of contaminating protein.
However, we can eliminate detergent in the affinity step and the
protein will remain soluble, although the yield will be reduced.
Eluents from the gel filtration column contained a potent activity
which selectively activated phospholipase A_2, although it had
absolutely no effect on phospholipase C activity utilizing either
phosphatidylcholine or phosphatidylinositol as substrate (Table
4).

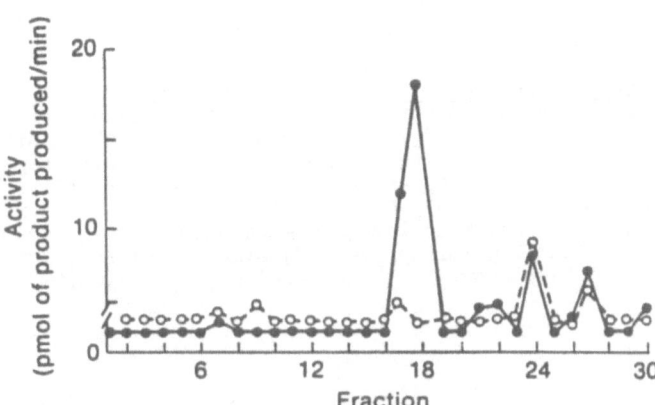

Figure 9. Affinity Chromatography of PLAP Affinity-purified anti-mellitin
antibodies were immobilized on high performance affinity columns. CPAE
cell-free sonicates were then loaded onto the column and the column was washed
extensively prior to elution with pH 3.1 sodium acetate buffer. Fractions were
collected and aliquots assayed in triplicate for stimulatory activity by mixing
cellular sonicates containing endogenous phospholipase A_2, substrate and 100ul
of column eluent for 60 min at 37°C. Results are expressed as pmol hydrolyzed
above control values (i.e., cell sonicate + substrate alone) which were 4
pmol/min.

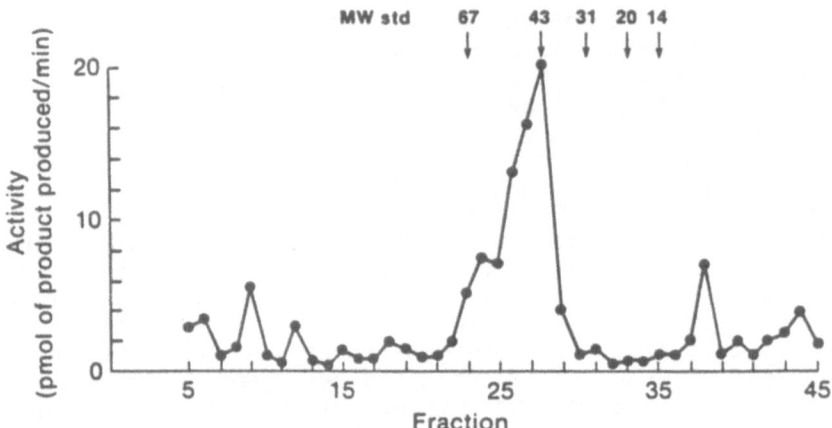

Figure 10. Size Exclusion Chromatography of PLAP The peak of the
stimulatory activity obtained from the affinity column was further purified by
size exclusion HPLC. Aliquots of each fraction were assayed in triplicate for
stimulatory activity.

TABLE 4

Activity	Phospholipase C		PLA	
	PC	PI	PC	PE
Sonicate only	4.9 + 0.6	2.1 + 1.1	8.2 + 0.9	2.6 + 0.6
PLAP only	.1	.1	.1	.1
Sonicate + PLAP	3.6 + 0.7	1.6 + 0.1	46.0 + 0.3	2.1 + 0.9

Enzyme and substrate specificity of PLAP activity The effects of 10 units
of PLAP on phospholipase A_2 and C activity were assayed using either
phosphatidylcholine (PC) phosphatidylethanolamine (PE), or phosphatidylinositol
(PI) as substrate. Results shown are the mean + S.D. for three separate
experiments performed in triplicate and are expressed as pmol of reaction
product produced/min/mg of protein. For experimental details see reference 23.
The substrates used were 1-palymitoyl-2-[^{14}C]-arachidonyl-sn-glycero-3-[choline
-methyl-^{3}H]-phosphocholine,1-palmitoyl-2-[^{14}C]-arachidonyl-sn-glycero-3-phospho
-[^{3}H]-ethanolamine and 1-stearoyl-2-[^{14}C]-arachidonyl-sn-glycero-3-phospho-
[^{3}H]-myoinositol.

Attempts were made at visualizing this protein on a silver-
strained sodium dodecyl sulfate-polyacrylamide gel. Although the
sensitivity of the silver stain was such that approximately 10 ng
of standard proteins could be easily visualized, no protein bands
in the PLAP fractions were detected. Therefore, to assess the
molecular weight of the putative protein constituent, aliquots of
the active fraction from gel filtration chromatography were
iodinated utilizing ^{125}I-Na and chloramine T prior to SDS-PAGE
electrophoresis. As can be seen in Figure 11, a single

predominant protein band at 28 kD was present. The observed
difference between the apparent molecular weight at which PLAP
eluted from the gel filtration column and that visualized after
autoradiography of the iodinated protein is likely due to the
presence of aggregates formed during gel filtration chromatography
since detergent was utilized to improve yields of PLAP from this
column (11).

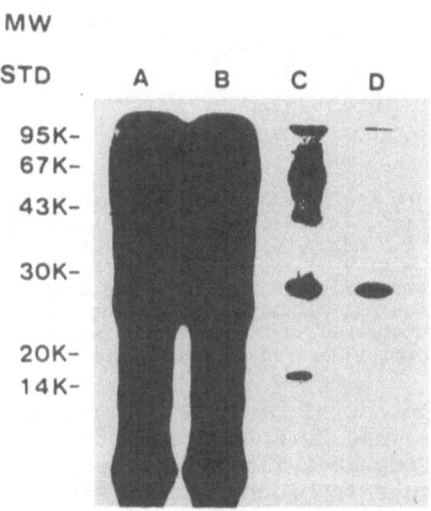

Figure 11. Polyacrylamide Gel Electrophoresis of PLAP. Polyacrylamide gel
electrophoresis was used to analyze the results from the protein purification.
Samples were iodinated using ^{125}I-Na and chloramine T, and eletrophoresed on
12% acrylamide gels. Identical results were obtained using Bolton-Hunter
reagent (data not shown).

To directly assess the mechanism responsible for PLAP
activation of phospholipase A_2, kinetic experiments were performed
and analyzed utilizing Lineweaver-Burk plots. As can be seen in
Figure 12 (11), treatment of phospholipase with PLAP resulted in a
five-fold increase in V_{max} although no apparent alteration in
substrate affinity for phospholipase A_2 was present. Although the
CPAE cytosol contained endogenous lipid, significant isotope
dilution was present only for the lowest substrate concentration
utilized. Furthermore, comparisons of the activation of
phospholipase activity at each substrate concentration with
varying amounts of PLAP are qualitatively valid since equivalent
amounts of isotope dilution occurred at equivalent substrate
concentrations and purified PLAP is lipid-free. Taken together,
these results demonstrate that the B3CH$_1$ and CPAE cell lines
contain a protein constituent with an apparent molecular weight of
28 kD which cross-reacts with anti-mellitin antibodies and is a
potent activator of phospholipase A_2.

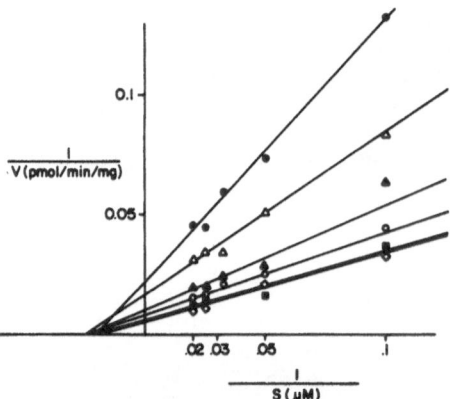

Figure 12. <u>Lineweaver Burk Analysis of Phospholipase A$_2$ Activity as a</u>
<u>Function of PLAP Concentration.</u> Increasing concentrations of PLAP in units: 0
units, K$_m$ = 49 V$_{max}$ = 45; 0.5 units, K$_m$ = 33 V$_{max}$ = 64; 1.0 units, K$_m$ = 45
V$_{max}$ = 106; 1.5 units, K$_m$ = 41 V$_{max}$ 131; 4.0 units K$_m$ 50 V$_{max}$ 222; 6.0
units, K$_m$ = 64 V$_{max}$ = 192 were added to the cell-free sonicate (1mg/ml), and
phospholipase A$_2$ activity was assayed using increasing concentrations of
phosphatidylcholine as a substrate. The data shown represent results from two
experiments with each data point assayed in triplicate.

        To identify the potential biologic significance of PLAP,
additional experiments were performed to determine: 1) if PLAP is
produced in humans during an inflammatory response; 2) if PLAP
produced in human tissues is biochemically similar to that
obtained from the CPAE cell line; 3) to identify the cellular
constituents in inflammatory regions which produce PLAP <u>in vivo</u>;
and 4) to determine if PLAP can initiate an inflammatory response
in mammalian tissues <u>in vivo</u>. Rheumatoid arthritis was chosen as
the prototypical human inflammatory disease mediated by enhanced
eicosanoid synthesis. In initial experiments, we demonstrated
that synovial fluid taken from patients with rheumatoid arthritis
contained PLAP as determined by dot blot analysis utilizing
anti-mellitin antibodies (Table 5). Furthermore, joint fluid from
a noninflammatory arthritic, osteoarthritic patients, contained
substantially less PLAP (p < .001) (Table 5). To determine the
cellular source of PLAP in the synovial fluid, joints removed from
patients undergoing joint replacement surgery were fixed,
sectioned and treated with anti-mellitin antibody followed by
immunolocalization procedures. These experiments demonstrated
that the major sources of PLAP in human synovial fluid were
macrophages and endothelial cells (Figure 13). Subsequently, we
purified PLAP from human synovial fluid by sequential immunoaffinity and gel filtration chromatographies and demonstrated that
the molecular mass of human PLAP was also approximately 28 kD.
Human PLAP obtained from these column chromatographic steps also
specifically activated phospholipase A$_2$ without demonstrable
effects on phospholipase C activity.   These results indicate that
human PLAP is biochemically similar to PLAP isolated from bovine
and murine sources.

TABLE 5

Arthritis Disease	N	Blot Reading
Rheumatoid Arthritis	31	1.49 + 1.09*
Osteoarthritis	18	0.32 + 0.46
Seronegative Spondyloarthropathy	5	0.36 + 0.41

*p < 0.001 by students t-test. Dots were rated 0-3 + (positive).

Immuno-Dotblots of Human Synovial Fluid PLAP The presence of a PLAP in synovial fluid from patients with rheumatoid arthritis has been determined in synovial fluids from patients with known rheumatoid arthritis, osteoarthritis (degenerative arthritis, a noninflammatory arthritis) and seronegative spondyloarthropathies (e.g., ankylosing spondylitis and Reiter's Syndrome (another type of noninflammatory arthritis). The presence of PLAP in synovial fluids was determined by immuno-dotblotting with rabbit antimellitin antibody in a double blind manner. The dotblots were rated 0-3+ (positive). Synovial fluid from rheumatoid patients contained significantly more PLAP than did fluid from the two control groups.

To determine if PLAP alone could initiate an inflammatory response, purified human PLAP was injected into the joint capsules of rabbits. Briefly, female New Zealand white rabbits (3 - 4 kg) were anesthetized and 1 ml of sterile pyrogen-free saline containing PLAP was injected into the knee joint. The contra- lateral knee was injected with an equal volume of either pyrogen-free saline alone, saline containing 0.15 ng of lipopolysaccharide (the maximal amount of LPS found in the PLAP fractions), saline containing 10 ug of actin (protein control), or

1 ml of eluent from the HPLC gel filtration column not containing PLAP activity. Injections were made through the joint capsule to avoid the infrapatellar fat pad. Twenty-four hours later the rabbits were necropsied and both knees were lavaged with 2 ml of sterile pyrogen-free saline. The lavage fluid was cultured for bacteria and the total and differential cell count determined. The synovial-lined intrapatellar fat pad portion from each joint was excised and fixed in formalin and embedded in paraffin prior to processing for histology (Figure 14). Bacterial cultures of synovial lavage fluids were negative. The dose-response relation- ship of PLAP as an initiator of cellular infiltration and joint destruction was determined. Twenty-four hours after injection of $3 - 50 \times 10^3$ units of PLAP, the mean total cell count of synovial lavage fluid from injected joints ranged from $3.2 - 5.1 \times 10^6$ cells/ml and was significantly increased compared to each control group. Although the inflammatory cells in the lavage fluid were more than 95% polymorphonuclear leukocytes (PMN), there was an absolute increase in both PMN and mononuclear inflammatory cells after PLAP injection. Lavage fluid obtained 24 hours after injection of PLAP contained significantly fewer PMN and more monocytes. Histologically, a dose-dependent inflammatory cell infiltrate of the infrapatellar fat pad was present beneath and

within the synovial lining after PLAP injection. Furthermore,
6000 units of PLAP induced a three fold higher synovial exudate
than the undiluted adjacent HPLC column fraction control which was
16 times more concentrated. Although the protein content of the
injected material was too small to measure, the fact that the
lowest concentration of PLAP tested (3000 units, 1 nanogram)
induced a 2- to 4-fold increase in the mean number of inflammatory
cells recovered from the synovial cavity demonstrates that PLAP is
a potent inflammatory stimulator in mammalian tissues.

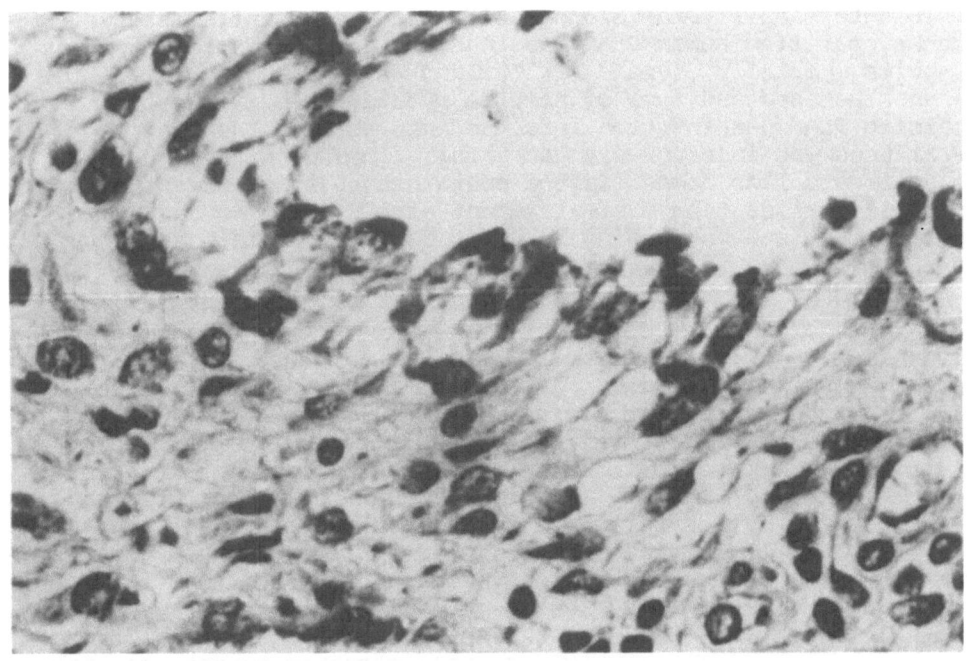

Figure 13. Immunohistologic Localization of PLAP in Human Rheumatoid
Synovium. Human synovial tissue was obtained from patients undergoing joint
replacement therapy. The tissue was fixed and prepared from
immunohistochemistry. Rabbit anti-mellitin antibody was used to localize PLAP
in this tissue. PLAP was found to be primarily localized in macrophages,
vascular smooth muscle and endothelial cells. Lymphocytes, synovial
fibroblasts and some synovial lining cells did not contain PLAP as determined
by this technique.

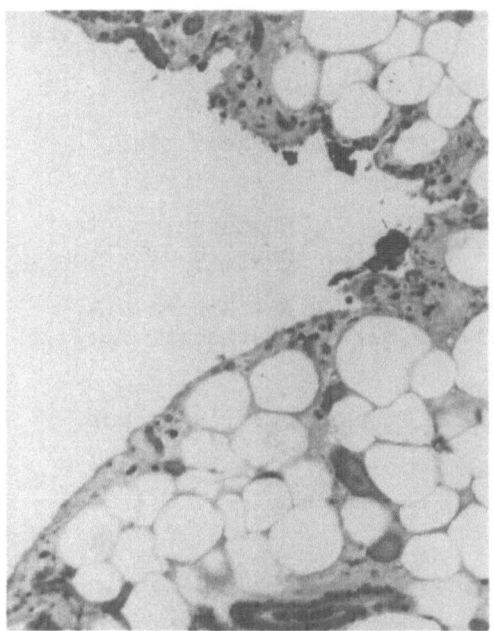 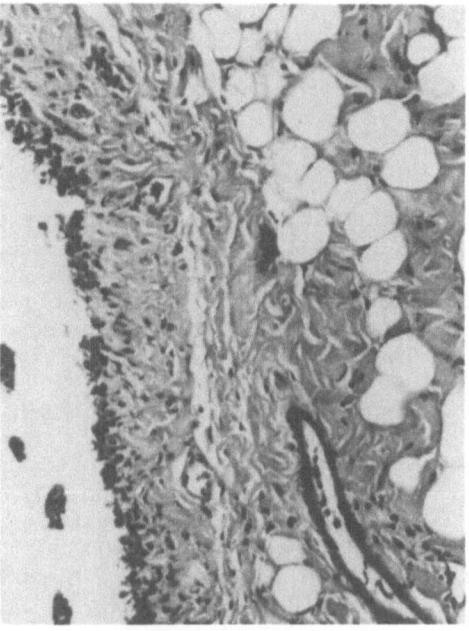

Figure 14. Effects of PLAP Injection Into Rabbit Knee Joints.
Infrapatellar fat pad from a rabbit necropsied 25 hours after the
intra-articular injection of 50,000 units of PLAP is shown. There is
widespread vasodilatation and diffuse inflammatory cell infiltrate (Left).
   An aliquot of HPLC eluent from the size exclusion chromatogram that did not
have PLAP activity was injected into the contralateral knee joint (Right). A
few inflammatory cells were observed near the synovial lining.

        Taken together, these results demonstrate that mammalian
tissues contain a 28 kD protein which is induced by LTD_4 and is a
potent activator of mammalian phospholipase A_2 activity. This
protein is present in increased amounts in human joint fluids
obtained from rheumatoid arthritic patients and, by itself, can
induce inflammation in mammalian joint tissues. Since the time
course of the accumulation of this protein parallels the
production of biologically active eicosanoids, and it is present
in increased amounts in the synovial fluid taken from inflammed
joints and since by itself, it can initiate an inflammatory
response in mammals, we believe that PLAP is an important
contributor to the initiation and propagation of the inflammatory
response in humans.

Reference

1. Brocklehurst, W.E. (1960) J. Physiol 151 416-423.

2. Drazen, J. and Austen, K.F. (1974) Proc. Natl. Acad. Sci. USA 77 4354-4359.

3. Hedqvist, P., Dahlen, S.E., Gustafsson, L., Hammarstrom, S. and Samuelson, B. (1980) Acta. Physiol. S. and. 110 311-319.

4. Weichmen, B.M., Muccitelli, R.M., Osborn, R.R., Holden D.A. Gleason, J.G. and Wasserman, M.A. (1982) J. Pharmacol. Exp. Ther. 222 202-218.

5. Clark, M.A., Cook, M., Mong, S. and Crooke S.T. (1985) Eur. J. Pharm 116 207-220.

6. Clark, M.A., Littlejohn, D., Mong, S. and Crooke S.T. (1986) Prostaglendins 31 157-166.

7. Clark, M.A., Conway, T.M. and Crooke, S.T. (1987) J Liquid Chrom, 10 2707-2719.

8. Clark, M.A., Conway, T.M., Bennett, C.F., Crooke, S.T. and Stadel, J.M. (1986) Proc. Natl. Acad. Sci. USA. 83 7320-7324.

9. Flower, R.J. and Blackwell G.J. (1976) Biochem. Pharmacol 25 285-291.

10. Clark, M.A., Conway, T.M. Mongs, Stiemen S. and Crooke S.T. (1981) J. Biol Chem 261 10713-10718.

11. Clark, M.A., Conway, T.M., Shorr, R.G.L., and Crooke, S.T. (1987) J. Biol. Chem. 262 4402-4406.

CELLULAR AND EXTRACELLULAR PHOSPHOLIPASE A_2 ACTIVITY

IN ZYMOSAN PLEURISY IN RAT

D.W. Morgan[*], C. Anderson, K. Meyers,
J. Coffey, K. Moody and A. Welton

Department of Allergy and Inflammation
Hoffmann-LaRoche, Inc.
Nutley, NJ 07110

SUMMARY

The pleural exudate from rats treated intrapleurally with zymosan contains phospholipase A_2 (PLA_2) activity which is Ca^{2+}-independent and optimally active at a neutral pH. This PLA_2 activity was found in approximately equal amounts in both the cellular and extracellular fractions of the exudate. The Ca^{2+}-independency of the PLA_2's in the pleural exudate distinguishes them from plasma PLA_2's and this suggests that the source of the exudate PLA_2's is not plasma. The appearance of PLA_2 activity in zymosan-induced pleural exudate correlates temporally with increases in exudate volume and pleural cell number. In all cases, the maximum response was seen 24 hr after zymosan challenge. All parameters of pleurisy and PLA_2 activity are similarly sensitive to the steroid dexamethasone which has been hypothesized to act, in part, through the synthesis of PLA_2 inhibitory peptides. In its entirety, this information suggests that there is a relationship between pleural PLA_2 activity and the appearance of pleural inflammation (exudate volume and cells) and that PLA_2 may play an important role in the initiation and propagation of this inflammatory process in rats. Furthermore, the zymosan-induced pleurisy model may serve as a useful model for the identification of PLA_2 inhibitors with antiinflammatory activity.

INTRODUCTION

Cellular and extracellular phospholipase A_2's (PLA_2's) are enzymes associated with inflammatory processes in experimental animals and in man (1). Zymosan-induced pleurisy in rodents is

---

[*]To who correspondence should be sent.

Phospholipase A2
Edited by P.Y.-K. Wong and E. A. Dennis
Plenum Press, New York, 1990

characterized by the accumulation of a pleural exudate fluid containing inflammatory cells (2-4) and has been used recently as an in vivo model for the evaluation of potential antiinflammatory agents. We have studied the characteristics of the PLA_2 activities in zymosan-induced, pleural exudate fluid in rat to evaluate the relationship of the activities of these enzymes to the development of this experimentally-induced inflammatory reaction.

MATERIALS AND METHODS

Zymosan-Induced Pleurisy

Zymosan (Sigma Chemical, Co., St. Louis, MO) in saline (10 mg/ml) was heated for 1/2 hr in a boiling water bath, washed three times and then resuspended in saline (10 mg/ml). This preparation of boiled zymosan did not contain detectable levels of PLA_2 activity. Male Lewis rats (260-300g) were given intrapleural injections of the 10 mg/ml zymosan suspension (0.25-2.0 mg zymosan/rat) to induce the pleurisy response. At the indicated times, the pleural cavity was exposed and heparin ($\approx$ 20 units/ml) was added to the exudate fluid in the pleural cavity to prevent coagulation. The exudate fluid was collected by aspiration and the volume was measured. An aliquot (10 μl) was removed for quantitating total cell accumulation using a Coulter Counter. The exudate fluid was then centrifuged at 10,000 x g for 30 sec.. The supernatant (extracellular fraction) was aspirated and the pellet (cellular fraction) was resuspended in a volume of saline equal to the original exudate fluid volume. All samples were stored frozen and thawed immediately before sonication and assay for PLA_2 activity. The cellular fraction, when thawed, was sonicated (Ultrasonic Cell Disrupter, Heatsystems - Ultrasonics, Inc.) 4 times for 5 sec. intervals prior to use.

PLA_2 Assay

The activity of PLA_2 was quantitated by the method of Vadas et al. (5). The standard reaction mixture (0.5 ml) contained 6.6 mg/ml fatty acid-free bovine serum albumin (Sigma Chemical Co. St. Louis, MO), 5.0 mM $CaCl_2$, 50 mM HEPES buffer (pH 7.4), 125 mM NaCl and [^{14}C]oleate-labeled E. coli (heat-killed, intact) as substrate (2 x 10^9 cells/ml which contained 0.48 μCi radioactivity/μmol phospholipid and 18.2 nmol total phospholipids/ml) (Dr. R. Franson, Medical College of Virginia, Richmond, VA). The reaction was terminated after 30 min. at 37° C by the addition of 0.5 ml 50 mM EDTA. The reaction mixture/EDTA was then filtered through a 0.45 μ Millex-HV filter (Millipore Corp.) to separate the released oleic acid from E. coli. The quantity of [^{14}C]oleic acid in the filtrate was determined by liquid scintillation spectrometry (Beckmann LS 7800 spectrometer). The efficiency of recovery of [^{14}C]oleic acid through the filtration step was 78% ± 4% and 1000 dpm was equivalent to 2.4 nmoles oleate hydrolyzed/hr (corrected for recovery). Heparin at 20 units/ml was not inhibitory to the PLA$_2$ activities in these studies.

RESULTS AND DISCUSSION

Time course of zymosan-induced pleurisy

As illustrated in Fig. 1., intrapleural injection of zymosan into rats caused a time-dependent increase in the volume of pleural exudate fluid and in the number of cells found in this fluid. Both

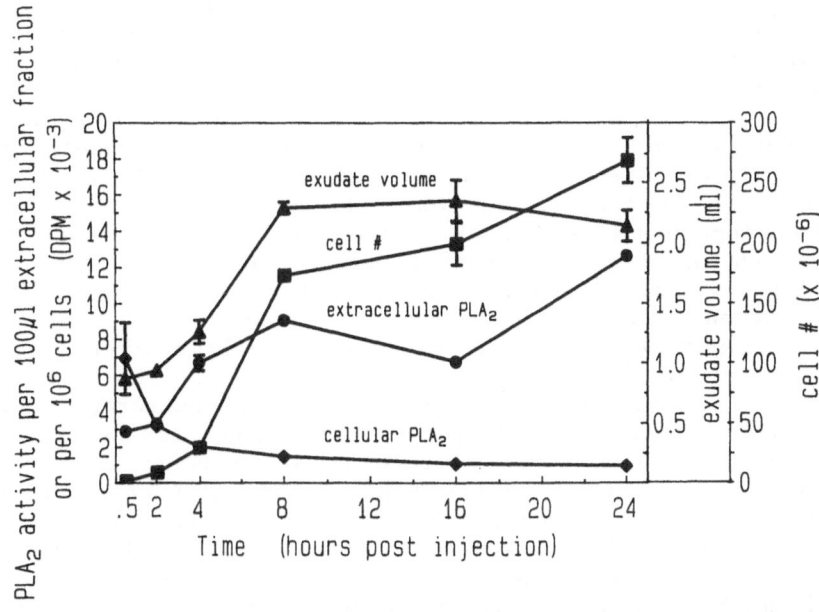

Fig. 1. Time-course for zymosan-induced pleurisy in rats. Zymosan was prepared and injected intrapleurally (2 mg/rat; n=8 for each time point). The pleural exudate was collected at various times (0.5-24 hr) and the volume and cell number determined. The exudate was separated into the cellular and extracellular fractions and the PLA$_2$ activity in each fraction determined (see Methods for details). ▲, exudate volume (ml); ■, cell number; ●, extracellular PLA$_2$ activity (dpm/100 μl exudate volume); ◆ , cellular PLA$_2$ activity, (dpm/10^6 cells). Values are ± SEM. For comparison, at the 4 hr time point, total extracellular PLA$_2$ activity = 8.4 x 10^4 dpm and total cellular PLA$_2$ activity = 5.9 x 10^4 dpm. Intrapleural injection of saline (200 μl) into untreated rats (controls) resulted after 30 min. in recovery of 110 μl of pleural fluid which contained 3.6 x 10^6 cells, 1.1 x 10^3 dpm/100 μl PLA$_2$ in the extracellular fraction and 0.28 x 10^3 dpm/10^6 cells PLA$_2$ activity in the cellular PLA$_2$ activity.

parameters increased dramatically between 0.5 and 8 hr during which time the cell number increased by more than 10-fold and the exudate volume increased by more than 2-fold. Between 8 and 24 hr, changes in all parameters were less dramatic than between 0.5 and 8 hr and the values were significantly lower at 48 hr than at 24 hr (data not shown). At 4 hr, the population of cells consisted primarily of neutrophils (>94%) and a few macrophages ($\approx$ 6%); whereas, at 24 hr, the proportion of macrophages increased ($\approx$ 34%) in comparison with the proportion of neutrophils (66%). Thus, intrapleural zymosan induced a classic inflammatory response which, over time, was similar to the pleurisy induced by carrageenan (6).

We observed that the pleural exudate also contained PLA_2 activity. PLA_2 increased dramatically in the extracellular fraction of the pleural exudate between 2 and 8 hr when expressed either as activity per unit volume ($\approx$2.5-fold) as indicated in Fig. 1. or as total activity ($\approx$5-fold). When expressed on the basis of total activity, PLA_2 in the cellular fraction of the pleural exudate also increased during the same time period ($\approx$10-fold as a result of a large increase in cell number); however, when expressed on a per cell basis (Fig. 1.), the activity of this cell-associated PLA_2 activity decreased during this time period. These data suggest either that the cells which infiltrated the pleural cavity during the progression of the pleurisy reaction (e.g. neutrophils observed at 4 hr.) contained decreased amounts of PLA_2 as compared with the cells (resident macrophages) which were present at earlier times or that the infiltrating cells released a portion of their PLA_2 into the exudate fluid, perhaps by exocytosis, as they encountered inflammatory mediators in the pleural space. Macrophages and neutrophils are known to contain PLA_2 and secrete it in response to various stimuli (1). While it is likely that the major portion of the soluble PLA_2 in the pleural cavity is released from the infiltrating cells, it is also possible, however, that the extracellular PLA_2 may be derived from the plasma where it is also found abundantly. That is, the plasma PLA_2 may move into the pleural cavity from the plasma (as do other plasma proteins) as a result of the increased vascular permeability associated with the pleurisy reaction. It seems unlikely that a major portion of the soluble PLA_2 in the pleural cavity is derived from plasma since the Ca^{2+} and pH requirements of plasma and pleural fluid PLA_2's differ significantly.

Calcium-dependency and pH-profiles of pleural and plasma PLA_2

The PLA_2 activity in the extracellular fraction was inhibited by 20-30% in the presence of 3 to 5 mM EGTA, a specific chelator of Ca^{2+}, as was the PLA_2 activity in the cellular fraction of the pleural exudate. In contrast, the PLA_2 activity in the plasma from zymosan-treated or control rats was inhibited by approximately 60% (Fig. 2.). Thus, the pleural PLA_2's do not appear to be Ca^{2+}-dependent. These results suggest that the exudate PLA_2's are different from the plasma

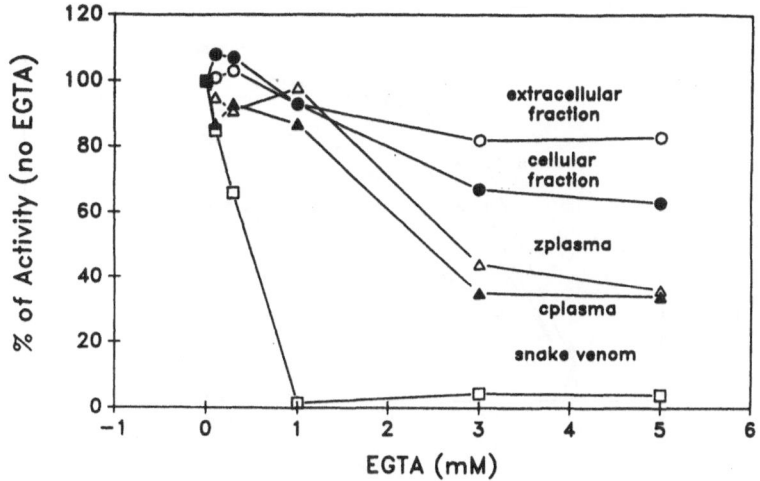

Fig. 2. Effect of EGTA on PLA$_2$ activities.   Rats were treated intrapleurally with zymosan (2 mg/rat, n=6).   After 4 hr, extracellular and cellular PLA$_2$ fractions were prepared from the pleural exudate as described in Fig. 1., samples of blood were taken, treated with heparin (20 units/ml) and the plasma separated by centrifugation.   PLA$_2$ activity was assayed as described in Methods except that 1.0 mM Ca^{2+} was included in the reaction mixture. Results were expressed as a percent of the activity measured in the reaction mixture containing 1.0 mM Ca^{2+} and no EGTA.   o, extracellular; ●, cellular; Δ, plasma from zymosan-treated rats (zplasma); ▲, plasma from untreated rats (cplasma); □, PLA$_2$ from Naja naja snake venom (Sigma Chemical Co., St. Louis, MO.)

PLA$_2$'s and are consistent with the hypothesis that platelets are a major source of plasma PLA$_2$.   Rat platelets are known to contain a high level of a Ca^{2+}-dependent PLA$_2$ (7).   In addition, pH versus activity curves for extracellular pleural PLA$_2$ activity, cellular pleural PLA$_2$ activity, and plasma PLA$_2$ activity demonstrated that the pH profile of extracellular pleural PLA$_2$ resembles that of cellular pleural PLA$_2$ activity and had a pH optimum between 6.0 and 7.5 (Fig. 3.). In comparison, no definitive pH optimum could be demonstrated for plasma PLA$_2$ activity. This may be due to the presence of a variety of PLA$_2$'s with a wide range of pH optima in plasma.

The lung is another possible source of the soluble PLA$_2$'s in the pleural exudate.   The cytosolic fraction of rat lung cells has been reported to contain a high amount of soluble, Ca^{2+}-independent PLA$_2$

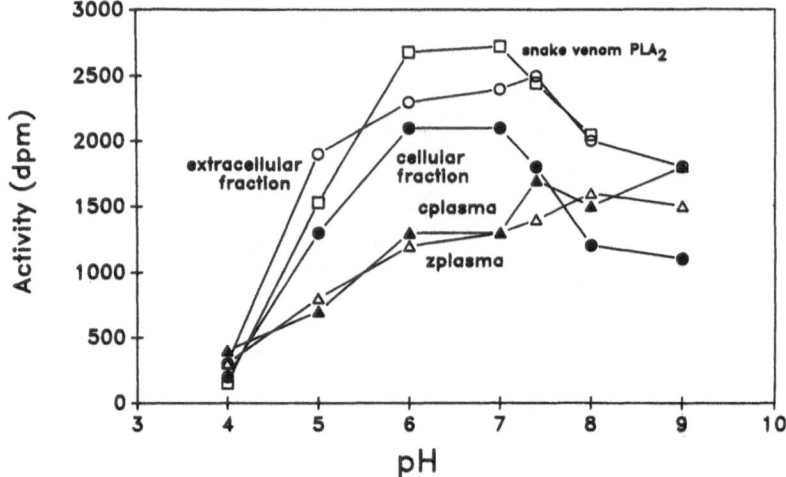

Fig. 3. pH versus PLA$_2$ activity relationship for PLA$_2$'s from various
sources. Samples for measurement of PLA$_2$ activity at various pH's
were obtained as described in Fig. 2. PLA$_2$ activity was assayed as
described in Methods except that the following buffer systems were
used in the assay mixture and adjusted to the desired pH: Na$^+$
acetate, pH 4-5; HEPES, pH 6-7; glycine-HCl, pH 8-9. O, extracellular; ●, cellular; Δ, plasma from zymosan-treated rats (zplasma);
▲, plasma from untreated rats (cplasma); □, PLA$_2$ from Naja naja snake
venom (Sigma Chemical Co., St. Louis, MO.).

relative to the amount of this enzyme in other rat tissues, including
blood (8).

Effect of zymosan dose on the development of zymosan-induced pleurisy

     To aid in the selection of optimal conditions for measurement of
effects of antiinflammatory agents, the dose-dependency of the
pleurisy response and PLA$_2$ activity to increasing concentrations of
intrapleural zymosan were measured at 4 hr (Fig. 4a.) and at 24 hr
(Fig. 4b.). At 24 hr, the exudate volume and the cell number
increased with increasing zymosan and were markedly greater than the
responses of these parameters to zymosan at 4 hr. At the 4 hr time
point, the total amount of cellular and extracellular PLA$_2$ activity
measured in the pleural exudate increased in a dose-related manner
with increasing zymosan at doses from 0.25 to 1.0 mg/rat zymosan
despite the fact that the total number of cells in the pleural cavity

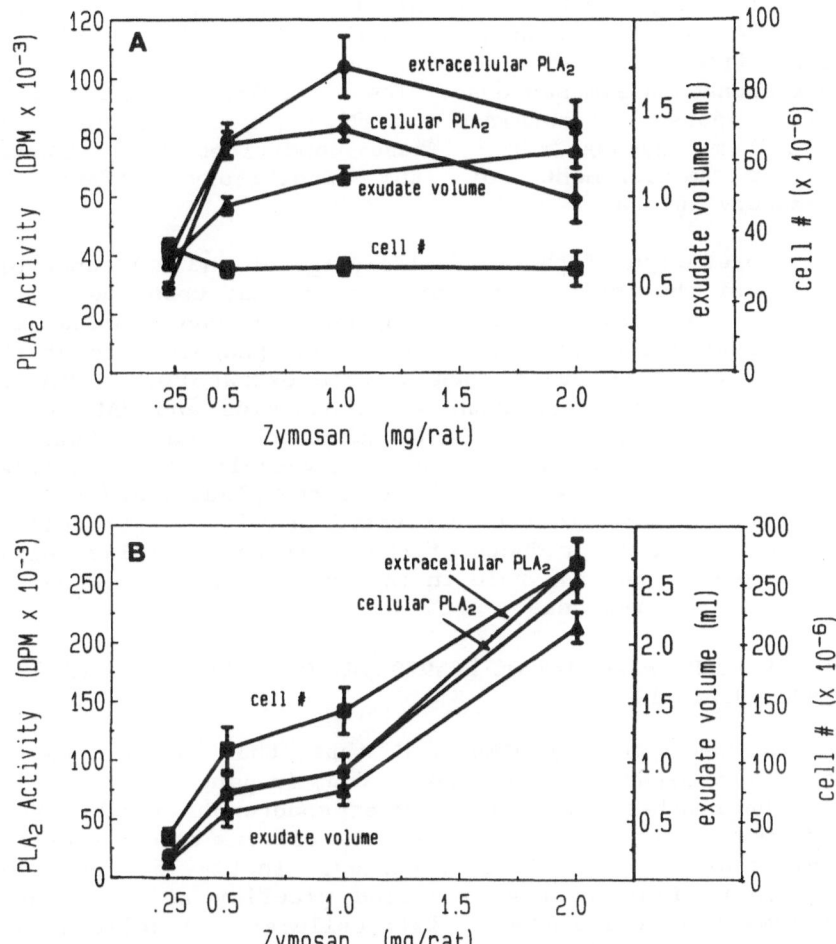

Figs. 4a. and 4b. Dose-response to zymosan for pleurisy and PLA_2 activities at 4 hr (Fig. 4a.) and 24 hr (Fig. 4b.). Zymosan was injected intrapleurally (0.25-2.0 mg/rat, n=8), the pleural exudate was collected and cellular and extracellular fractions were prepared as described in Fig. 1. and Methods. ▲, exudate volume (ml); ■, cell number; ●, total extracellular PLA_2 activity (dpm/rat); ◆, total cellular PLA_2 activity, (dpm/rat).

did not increase over this dose range. Therefore, the amount of PLA_2 activity per cell must have increased with higher doses of zymosan. At 24 hr, the total amount of cellular and extracellular PLA_2 increased in direct proportion to the number of cells and exudate volume, respectively, and was greatest at 2.0 mg zymosan/rat. However, at 24 hr, PLA_2 associated with either the cellular or extracellular fraction, when calculated on a per unit basis, did not change in response to zymosan dose. These results indicated that a the maximum response was obtained for all parameters 24 hr. after injecting 2.0 mg zymosan/rat. These conditions thus appeared suitable for measurement of the inhibitory effects of antiinflammatory agents.

One interpretation of these results (Fig. 4a and b.) concerning the mechanism of the inflammatory process is that upon the initial exposure of the pleural cavity to zymosan, resident inflammatory cells are activated through contact with, and phagocytosis of this irritant. These active cells then secrete extracellular PLA_2 and other inflammatory mediators such as eicosanoids and PAF through activation of cellular PLA_2. These substances induce immediate increases in the permeability of the pleural vasculature and initiate the recruitment of inflammatory cells into the pleural cavity. The newly recruited cells may also be activated to release extracellular PLA_2 and proinflammatory products of the activated cellular enzyme. Thus PLA_2 may play a central role in the initiation and propagation of this inflammatory response.

Effect of treatment with dexamethasone in the development of 24 hr zymosan-induced pleurisy

Since it has been hypothesized that the antiinflammatory properties of dexamethasone (dex) may be due, in part, to its ability to promote the synthesis and activity of endogenous inhibitors of PLA_2 (9), it was of interest to evaluate the effects of dex treatment on development of zymosan-induced pleurisy. In Fig. 5., dex (10 - 50 μg/kg, p.o.) significantly inhibited pleurisy as measured by exudate volume and cell number. Extracellular and cellular PLA_2 activities in the pleural exudate were also decreased by dex treatment in the 24 hr model. Similar results were obtained in the 4 hr model (data not shown) at the 50 μg/kg dose in which dex inhibited the accumulation of exudate fluid by 75%. The reason for the differences in sensitivity of the cell number to dex compared to the other measures taken is not apparent. Thus, because the inhibition of pleurisy and PLA_2 by dex treatment appear to correlate, these observations suggest a correlation between the level of PLA_2 activity and inflammatory response in zymosan-induced pleurisy.

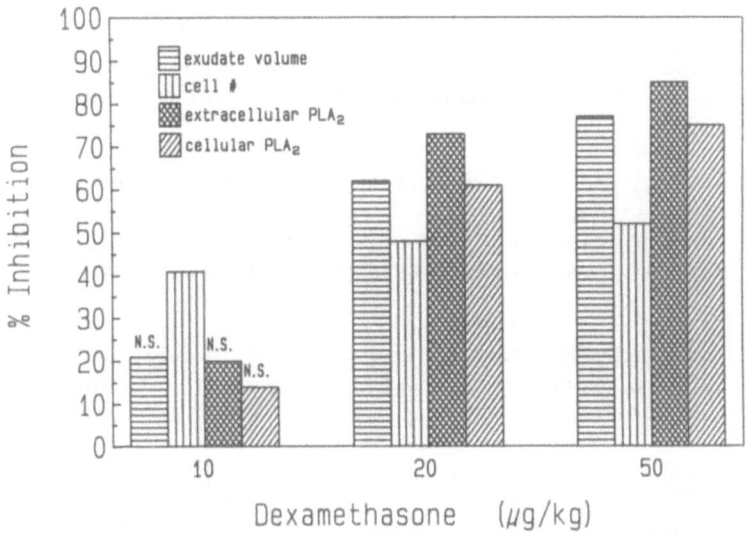

Fig. 5. Effects of dex on zymosan-induced pleurisy and PLA_2 activities 24 hr after zymosan injection. Dexamethasone was given orally in ASV one hr prior to injection of zymosan (2 mg/kg, n=6). Pleural exudate was collected and total volume and cell number determined. Extracellular and cellular fractions were prepared as described in Fig. 1. Total extracellular and cellular PLA_2 activities were calculated; % inhibition in dex treated group was calculated as 100 x [100 - (+)dex response ÷ (-)dex response]; n.s. = not significant, $P < 0.05$, Students t-test. ASV (aqueous suspending vehicle), NaCl (0.9%), carboxymethyl cellulose (5 g/L), benzyl alcohol (8.6 ml/L and Tween 80 (3.9 ml/L) in one liter distilled water.

CONCLUSION

We have found that the intrapleural treatment of rats with zymosan induces a pleural exudate fluid containing Ca^{2+}-independent PLA_2 activity which is optimally active at neutral pH. This PLA_2 activity correlates with the initiation, progression and dexamethasone-sensitivity of two parameters of pleurisy (exudate volume and cell number) and, thus, may serve as a useful model for the development of PLA_2 inhibitors with antiinflammatory activity.

REFERENCES

1. P. Vadas and W. Pruzanski, Biology of disease: role of secretory phospholipase A_2 in the pathobiology of disease. Lab. Invest. 55:391-404, 1986.

2. M. A. Martins, P. M. R. e Silva, H. C. Castro Faria Neto, P. T. Bozza, P. M. F. L. Dias, M. C. R. Lima, R. S. B. Cordeiro and B. B. Vergaftig. Pharmacological modulation of PAF-induced rat pleurisy and its role in inflammation by zymosan. Br. J. Pharmacol. 96:363-371, 1989.

3. K. Meyers, D. W. Morgan, C. Anderson, K. Moody, J. Coffey, and A. Welton. Effect of antiinflammatory compounds on phospholipase A_2 activity in zymosan-induced rat pleurisy. FASEB Journal 3:A908 (1989).

4. J. P. Tarayre, A. Delhon, M. Aliaga, M. Barbara, F. Bruniquel, V. Caillol, L. Puech, N. Consul and J. Tisne-Versailles. Pharmacological studies on zymosan inflammation in rats and mice. 2: Zymosan-induced pleurisy in rats. Pharmacological Research 21:385-395, 1989.

5. P. Vadas, E. Stefanski, and W. Pruzanski. Influence of plasma protein on activity of proinflammatory enzyme phospholipase A_2. Inflammation 10:183-193, 1986.

6. N. Ackerman, A. Tomolonis, L. Miram, J. Kheifts, S. Martinez and A. Carter. Three day pleural inflammation: A new model to detect drug effects on macrophage accumulation. J. Pharmacol. Exp. Ther. 215:588-595, 1980.

7. K. Horigome, M. Hayakawa, K. Inoue, and S. Nojima. Selective release of phospholipase A_2 and lysophosphatidylserine-specific lysophospholipase from rat platelets. J. Biochem. 101:53-61, 1987.

8. A. J. Pierik, J. G. Nijssen, A. J. Aarsman and H. Van den Bosch. Calcium-independent phospholipase A_2 in rat tissue cytosols. Biochim. Biophys. Acta 962:345-353, 1988.

9. R. J. Flower. Lipocortin and the mechanism of action of the glucocorticoids. Br. J. Pharmacol. 94:987-1015, 1988.

EXTRACELLULAR PHOSPHOLIPASE A$_2$ ACTIVITY IN TWO IN VIVO MODELS OF INFLAMMATION

Kathleen R. Gans, Susan R. Lundy, Randine L. Dowling,
William M. Mackin, Theresa M. Stevens, and Janet S. Kerr

E. I. du Pont de Nemours & Co. (Inc.)
Medical Products Department
P.O. BOX 80400
Wilmington, DE 19880-0400

SUMMARY

Two "in vivo" models of inflammation have been used to investigate the
role of phospholipases A$_2$ (PLA$_2$) in inflammation. These models are casein-
induced peritonitis in the rat and zymosan-induced peritonitis in the mouse.
The extracellular PLA$_2$ activities from peritoneal lavage fluid in these two
models are similar: they are calcium dependent and have broad neutral pH
optima. However, the relationship between extracellular PLA$_2$ activity and
cell influx in these models are not identical. In zymosan peritonitis, PLA$_2$
activity preceeded peak cell influx, reaching a maximum within 15 min after
zymosan injection, while cell influx peaked by 8 hr. In casein-induced
peritonitis, the PLA$_2$ activity peaked at 24 hr, while cell influx continued
through 48 hr. The origins of the PLA$_2$ activities in both models remain
unclear; one potential source is the plasma. Understanding the role of
extracellular PLA$_2$ activity in "in vivo" models, and investigating specific
inhibitors in these models may aid in our understanding of the role of
extracellular PLA$_2$ in diseases such as rheumatoid arthritis, endotoxin shock
and pancreatitis.

INTRODUCTION

PLA$_2$ cleaves the sn-2 acyl linkage of phospholglycerides to yield the free
fatty acid, arachidonic acid (AA) and a lysophospholipid. In mammalian cells,
there are at least two classes of PLA$_2$, neutral-active ones which are usually
calcium-dependent, and calcium-independent lysosomal PLA$_2$, with acid pH
optima. In addition, PLA$_2$ can be found both intra- and extracellularly. The
intracellular enzyme(s) may be associated with the α-granules in rat
platelets[1] and the plasma membranes[2,3]. The origin of extracellular PLA$_2$
remains unclear. The focus of this review is on the characterization of the
extracellular PLA$_2$ activities found in two models of inflammation.

There are three lines of evidence suggesting that extracellular PLA_2 plays a role in inflammation. First, the injection of human non-pancreatic PLA_2 produces inflammation in experimental animals[4,5]. Second, elevated levels of extracellular PLA_2 activity are found in sera from patients in several disease states, including rheumatoid arthritis[6], endotoxin shock[7] and pancreatitis[8]. Third, extracellular PLA_2 activities increase in response to the injection of various inflammagens. In this review, we will focus on this last observation.

Two models which have been used to investigate the role of PLA_2 in vivo involve inflammatory exudates induced by casein in the rat[9,10,11], and by zymosan in the mouse[12,13,14]. Although both models have been previously described, we have characterized them further. The extracellular PLA_2 elicited by casein is similar to other extracellular PLA_2 in that it is calcium dependent and has a neutral pH optimum[15,16,17,18]. More importantly, this enzyme activity has characteristics associated with the cells involved in the inflammatory process; the casein-elicited PLA_2 has the same N-terminal sequence as the rat platelet-associated PLA_2 and may have a sequence similar to the PMN-associated PLA_2[1,9]. Extracellular PLA_2 from human synovial fluid of rheumatoid arthritis patients is not different from rat platelet PLA_2, as determined by amino acid composition and partial N-terminal amino acid sequence[19,20,21]. Since the PLA_2 activity from casein-induced inflammation is similar to human extracellular synovial enzyme, and since the rat enzyme is readily accessible, this model can serve as a useful source for PLA_2 activity.

In the second model, zymosan, a yeast cell wall carbohydrate, induces an inflammatory response when injected into the peritoneal cavity. This inflammation is characterized by writhing, edema, and leukocyte influx.[12,13] We have observed the appearance of extracellular PLA_2 activity and AA in the peritoneal cavities from zymosan-treated mice.[13] Because the appearance of extracellular PLA_2 activity could be a key event in the inflammatory process, we further investigated its characteristics by determining calcium dependence and pH optima, and related the appearance of PLA_2 in time to the appearance of other components of acute inflammation, including cell influx. The amino acid sequence of this PLA_2 remains to be determined.

MATERIALS AND METHODS

Casein-Induced Inflammation in Rats

Preparation of E.coli Substrate. The membrane phospholipids of E.coli (strain HB101) were labeled with ^3H-oleic acid (NEN) according to the method of Patriarca et al.[22] as modified by Davidson et al.[23] The radiolabeled E.coli were autoclaved at 120°C and 2.7 kg/cm^2 for 15 min. More than 90% of the incorporated label was in the 2-acyl position of membrane phospholipids as determined by treatment with PLA_2 from Crotalus adamanteus (Sigma Chemical Co.).

Extraction of E.coli Phospholipid and Determination of Specific Radioactivity. Radiolabeled E.coli cells were extracted according to the method of Bligh and Dyer[24]. The phospholipid obtained was resuspended in chloroform and quantitated on the basis of inorganic phosphate (Pi)[25]. The specific radioactivity of this lipid was 5000 DPM/nmol of Pi. In our culture system, 6.2×10^8 cells contained approximately 20 nmol of phospholipids; this agrees with previously reported values[26].

Preparation of PLA$_2$ from Casein-Treated Rats. Male Lew/Cr1Br rats (Charles River Breeding Laboratories) weighing 200-250 g were used. Cells and fluid were isolated according to the method of Mackin et al.[27]. Rats were injected intraperitoneally (i.p.) with 10 ml of 6% (w/v) sodium caseinate (Sigma Chemical Co.) or 0.9% saline for controls. Peritoneal cavities were lavaged with 10 ml heparinized (100 units) saline 16 hr after injection. The fluid was centrifuged at 500 X g for 10 min at 4°C. The supernatant was centrifuged again at 2000 X g. Using this method, approximately 15 ml of fluid per rat was obtained. The cell pellets were pooled and washed three times in Hank's buffer (pH 7.2) at 4°C. Cells obtained in this manner were more than 90% polymorphonuclear leukocytes (PMN) as determined by Wright stain and light microscopic examination. In time course experiments, peritoneal cavities were lavaged from 5.5 to 48 hr following casein injection.

Acid extraction of PLA$_2$ activity from cells and lavage fluid was performed according to the method of Fawzy et al.[28]. Equal volumes of cells or fluid and ice-cold 0.36 N H$_2$SO$_4$ and 1.6 M NaCl were mixed for 3 hr at 4°C. The mixtures were centrifuged at 20,000 X g for 20 min. The supernatants were dialyzed overnight against two changes of 10 mM sodium acetate, pH 4.5, and the dialysate was centrifuged at 20,000 X g for 20 min. The supernatants were designated "acid-extracted" cells or fluid.

Zymosan-Induced Peritonitis in Mice

Animals and Cells. Male CF1 mice (fasted overnight, 16-20 g, Charles River Breeding Laboratories) were injected i.p. with 1 mg zymosan A (Sigma Chemical Co.) prepared in 0.5 ml of 0.9% saline at a final concentration of 2 mg/ml. Prior to injection, the zymosan was bead-milled for 1 hr at 4°C. Control animals were similarly injected with 0.5 ml saline or were uninjected. At various times after injection, mouse peritoneal cavities were lavaged with either 4 ml saline containing either 1% w/v sodium heparin (Upjohn) or with 4 ml Hank's buffer (without Ca^{++} or Mg^{++}) (Gibco) for the PLA$_2$ assays. All samples were collected into polypropylene tubes and visibly bloody samples were discarded. Lavage fluids were then centrifuged at 400 X g for 10 min at 4°C and cell free supernatants and cell pellets were separated. Cell free supernatants were fast frozen in a dry ice/isopropanol slurry and stored under argon or nitrogen at -70°C. Cell free supernatants that were subsequently used for the analysis of lipids or AA were acidified to pH 3.0 with 1N HCl prior to freezing. Cells were characterized morphologically from Wright-stained slides.

Enumeration of Cell Numbers in Lavage Fluids. Zymosan has been previously shown to stimulate the formation of leukocyte aggregates[12,14]. For this reason, the number of cells present in the lavage fluids was determined by DNA content, described by the original method of Burton[32] as modified by Giles and Meyers[33]. For this analysis, the peritoneal lavage cell pellet was first resuspended in 1 ml cold distilled water and DNA was extracted with 2 ml 10% w/v trichloroacetic acid (Sigma Chemical Co.). The mixture was centrifuged at 1100 X g for 15 min at 4°C and the cell free supernatant discarded. The pellet was resuspended in 1 ml 1.0 M perchloric acid by vortexing. The solution was heated for 20 min in a 70°C waterbath, with a 1 min vortexing at 10 and 20 min into the incubation. The mixture was then cooled for 5 min on ice. Samples were then centrifuged (1100 X g, 15 min, 4°C) and a 0.5 ml aliquot was removed and mixed with 1 ml of modified Burton's reagent in 100 ml glacial acetic acid. Samples were then incubated at 37°C for 16-18 h in the dark, and optical density measured at 600 nm. Standard curves were generated using calf thymus DNA (Sigma Chemical Co.) dissolved in 1.0 M perchloric acid as the standard. Data were converted to µg DNA by extrapolation from the standard curve.

HPLC analysis. AA was isolated from lavage fluid using a C18 solid phase extraction procedure (3 ml, J.T. Baker). Columns were preconditioned with two 3 ml volumes of HPLC grade methanol followed by two 3 ml volumes of saline. After sample application to the column, the column was washed with 2.5 ml 20% acetone and the column then dried for 3 min under vacuum. AA was eluted from the column with 3 consecutive 1.0 ml washes of hexane/ethyl acetate/acetic acid (56.4:3:0.6). Eluates were combined and dried under nitrogen at 37°C and resuspended in acetonitrile for HPLC analysis.

HPLC analysis was performed on a Novapak C18 column (15 cm X 3.9 mm I.D., Waters) coupled to an ODS-5S (3 cm. X 4.6 mm I.D.) guard column (BioRad). The column temperature was 35°C, and the mobile phase was acetonitrile/H_2O-0.1% glacial acetic acid (80:20). The flow rate was 2 ml/min. Quantitation of AA in the samples was derived from the peak area at 200 nm and extrapolation from a standard curves constructed using authentic AA from Nuchek Prep (Elysian). Internal standards of ^{14}C-AA (0.15 ng) and unlabeled prostaglandin B_2 were included in each sample.

PLA$_2$ Assay using E.coli Substrate.

PLA$_2$ activities from both casein and zymosan-elicited lavage samples were assayed by a modified method of Rothhut et al.[29]. ^{3}H-oleic acid labeled E.coli (50 µl, 20 nmol Pi) was added to the lavage fluid (10 µg protein in 100 µl of 1:100 dilution of lavage fluid with 0.9% saline) in 100 mM HEPES (Gibco), pH 7.5 containing 0.2 mM calcium (standard PLA$_2$ assay buffer), in a final volume of 250 µl. After 60 min at 37°C in a shaking water bath, the samples were placed on ice. Hydrochloric acid (HCl) (100 µl of 2 N HCl) was added to stop the reaction and 100 µl of 2% bovine serum albumin (BSA) to trap the release ^{3}H-oleic acid. The tubes were vortexed, then centrifuged at 1000 X g for 10 min to pellet the bacteria. The supernatant, which contained the liberated fatty acids, was counted by liquid scintillation spectrometry (Hewlett-Packard 2200CA). To determine the percent recovery, the pelleted bacteria were hydrolyzed by adding 200 µl of 1 N NaOH, at 60°C for 30 min. HCl (200 µl, 1 N) was added to neutralize the NaOH and the fluid was counted as above. In similar experiments the pH, calcium, protein concentrations, and time of incubation at 37°C were varied. In experiments varying the pH, the PLA$_2$ activity was assayed as above with buffer substitutions of 100 mM sodium acetate, (pH 4-6), 100 mM Tris/maleate (pH 6-8), 100 mM HEPES (pH 7-9), or 100 mM glycine (pH 9-10). Protein levels were determined by the Bradford method[30] using BSA as the standard. Results were reported as an average of triplicate samples ± SE and expressed as nmol phospholipid (PL) hydrolyzed/mg protein/hr. This was calculated as percent of total radioactivity and assumed that 6.2 X 10^8 E.coli cells equaled 20 nmol phospholipid.

Statistical Analysis

Where appropriate, statistical analysis was performed using an unpaired Student's t-test[31]. A p value of $p \leq 0.05$ was considered to be statistically significant.

RESULTS

Characterization of PLA$_2$ Activities

Acid extraction of PLA$_2$ activity has been used by others as a crude method of enzyme purification. Specific activities of crude cell homogenates and lavage fluids were determined 16 hr after casein injection, a time when > 90% PMN are present in the cavity, and were compared with the specific activities of acid-extracts of the same samples (Table 1).

Little (5 nmol PL/mg protein/hr) PLA$_2$ activity was present in whole cell homogenate as compared to the PLA$_2$ activity from lavage fluid (246 nmol PL/mg protein/hr). PLA$_2$ activity in cell acid-extracts was 20 times higher (100 nmol/mg protein/hr) than in the whole cell homogenate. Surprisingly, PLA$_2$ activity in extracellular fluid was not enhanced by acid-extraction.

Table 1. PLA$_2$ Activity in Total Cells or Extracellular
          Fluid from Casein-Treated Rats.

(nmol PL hydrolyzed/mg protein/hr)

Lavage Fluid	Cells	Acid-Extracted Fluid	Acid-Extracted Cells
246 ± 40	4.5 ± 1.4	200 ± 1.0	100 ± 8.0*

Cells and extracellular fluid were isolated as described in Materials & Methods, 16 hr after 6% casein injection. Both the fluid and the cell sonicates were acid-extracted as described. PLA$_2$ activity was determined using E.coli as the substrate. Data are expressed as mean ± SEM (n = 4). *P < 0.05 compared with unextracted cell homogenates.

The characteristics of the PLA$_2$ activities in both models are shown in Fig. 1 and 2. They are similar, although the fluid and cell samples from zymosan treated mice were not acid extracted. The levels of PLA$_2$ activity in acid-extracts of cells from casein-injected rats increased from 0.016 nmol PL hydrolyzed/hr at 0.4 μg protein to 2.48 nmol PL hydrolyzed/hr at 320 μg protein. There was a three-fold higher PLA$_2$ activity in acid-extracts of fluid (6.1 nmol PL hydrolyzed/hr) at 320 μg protein than in acid-extracts of cells from casein-treated rats. Acid extraction of cells results in a 20-fold purification of PLA$_2$ activity[34]. PLA$_2$ activities were linear up to 10 μg protein in acid-extracts from both cells and lavage fluid. The PLA$_2$ activities from both the casein and zymosan-treated animals were maximal in the neutral pH range (pH 7-10; Fig. 1B and 2A). All activities were calcium-dependent, and EDTA or EGTA inhibited these activities (Fig. 1C and 2B). Using standard assay conditions, the maximum PLA$_2$ activity was between 191 and 200 nmol PL hydrolyzed/μg protein/hr from the acid-extracted fluid from the casein-treated animals, while it was between 20 and 40 nmol PL hydrolyzed/μg protein/hr unextracted lavage fluid from the zymosan-treated animals.

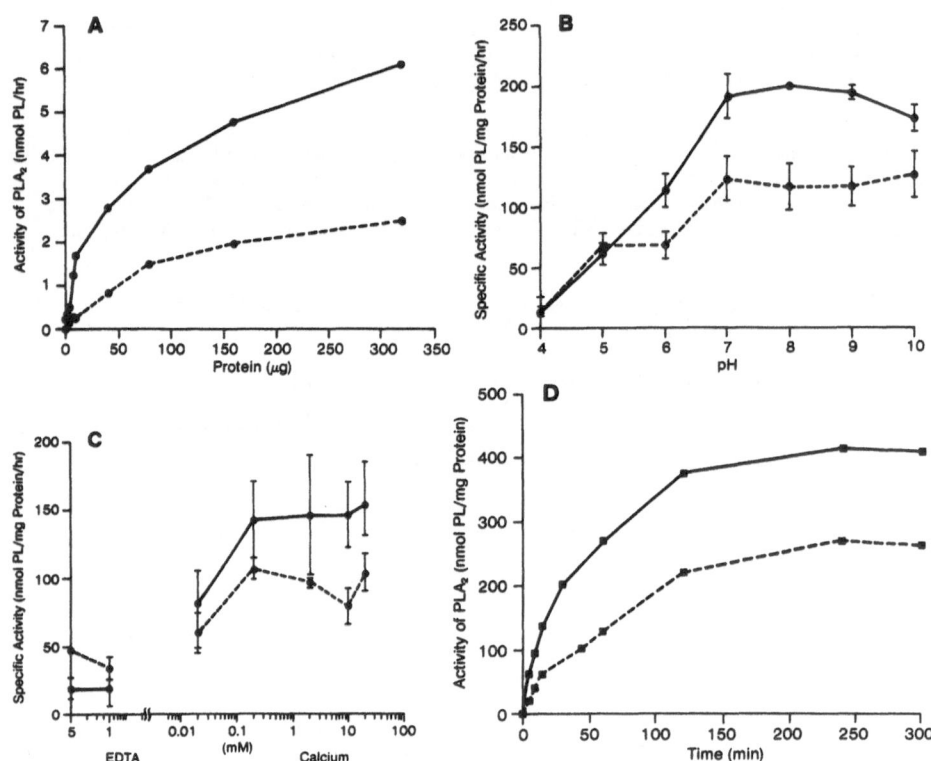

Fig. 1. Characteristics of PLA$_2$ activity in acid-extracts
of peritoneal lavage fluid (___) or cells (_ _ _) from
casein-treated rats using E.coli as substrate. (A.)
Protein (B.) pH (C.) Calcium (D.) Time of incubation.
Values are mean ± SE for two or three experiments.

Presence of AA, Cell Influx and PLA$_2$ Activity in Peritoneal Fluid from Zymosan-Treated Mice.

AA is one of the products of PLA$_2$ activation. We therefore chose to evaluate the kinetics of the appearance of PLA$_2$ activity and this product. Fig. 3 shows the time course of appearance of AA in the peritoneal cavities of mice after injection of 1 mg zymosan. Elevated levels of AA (4 fold increase

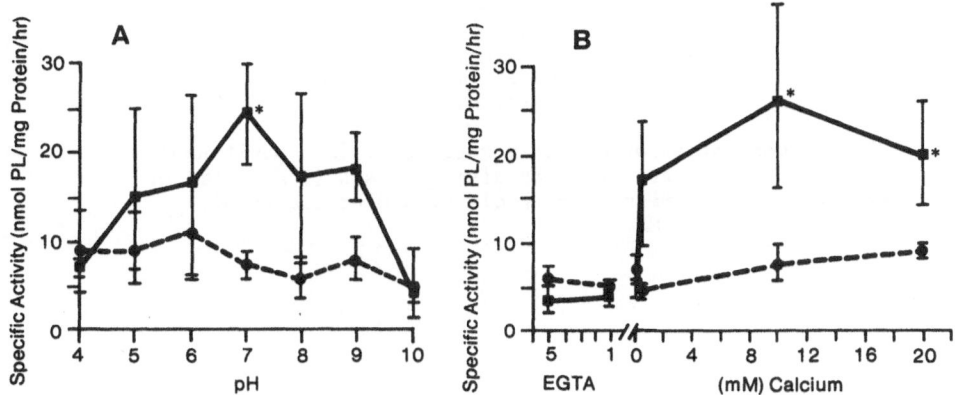

Fig. 2. Characteristics of PLA$_2$ activity in lavage fluids from zymosan-treated (■) or saline-treated (●) mice using E.coli as substrate. (A.) Calcium (B.) pH. Values are mean ± SE for two experiments.

over saline) were detected in cell free lavage fluids collected from mice 15 min (245 ± 34 ng/mouse) after zymosan injection and reached maximal 6 fold increase 1-2 hr after injection (388 ± 36 and 343 ± 24 ng/mouse). In contrast, there were no detectable changes in the AA levels measured in cell free lavage fluids from saline injected animals over the same time course.

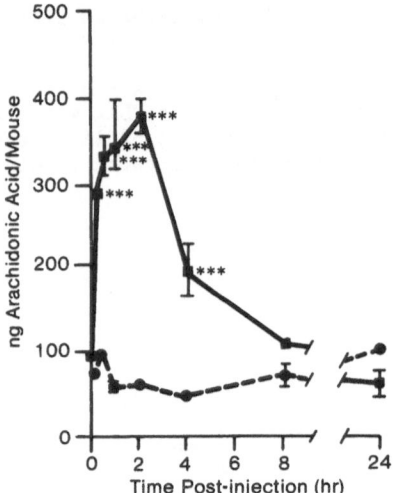

Fig. 3. Concentration of AA in mouse cell free peritoneal lavage fluid at various times after zymosan (1 mg) (■) or saline (●) injection. N=6 pools of 2 mice/timepoint. Values are mean ± SE.

In the zymosan peritonitis model, the influx of cells began 2 hr after zymosan injection and plateaued by 8 hr (Fig. 4A). By 4-6 hr after zymosan injection, all cells had phagocytized zymosan particles and could not be identified morphologically. Twenty-four hr after zymosan injection, the non-aggregated cell population was 50% PMN and 50% mononuclear cells. The cell population before zymosan injection was predominantly mononuclear (98%).

As shown in Fig. 4B, measurable PLA_2 activity was detected in the cell free lavage fluids within 5 min (2 fold increase, $p \leq 0.05$ vs. saline) after zymosan injection. Maximal PLA_2 activity was observed in the fluids from zymosan injected mice between 15 (32 ± 9 nmol PL hydrolyzed/mg protein/hr) and 30 min (28.2 ± 6.5 nmol PL hydrolyzed/mg protein/hr) after zymosan injection with a decrease occurring at later time points (6-24 hr). This activity was significantly higher ($p \leq 0.05$) than the PLA_2 activity detected 15 min and 30 min after saline injection (2.9 ± 10 and 3.5 ± 0.6 PL hydrolyzed/mg protein/hr, respectively). Thus, PMN infiltration occurred at a later time after zymosan injection than did the appearance of PLA_2 activity.

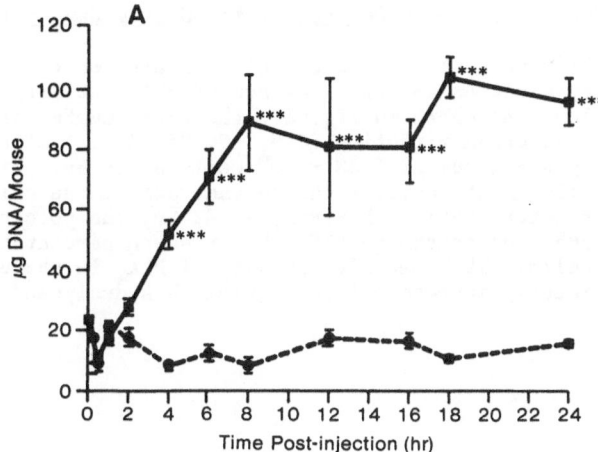

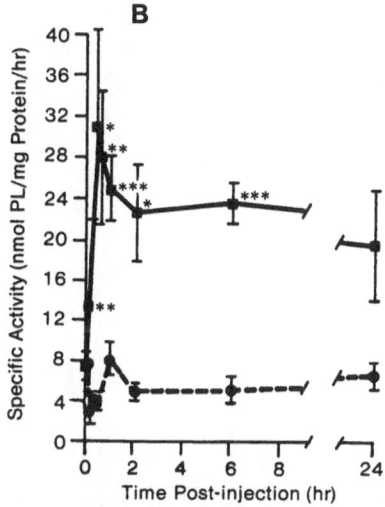

Fig. 4 (A). Time course of cell influx into the peritoneal cavity of mice after zymosan (■) or saline (●) injection as measured by DNA content. (B). PLA_2 activity in lavage fluid was measured. *$p \le 0.05$, ***$p \le 0.001$ compared to respective saline control, mean ± SE, 6 mice/group.

<u>Cell Influx and PLA₂ Activity in Peritoneal Fluid from Casein-Treated Rats.</u>

We and others[9],[10] have observed a correlation between the appearance of
PMN in inflammatory exudates and the presence of PLA_2 activity in casein-
induced inflammation. As shown in Fig. 5., the total number of cells in the
peritoneal cavity increased with time (5.5, 16, 24, 48 hr) after casein
injection, reaching a maximum of 1.33 x 10⁹ cells at 48 hr. Differential
analysis of the cells in the samples showed that 95% of the cells were PMN at
5.5, 16, and 24 hr after casein. However, at 48 hr, the percentage of total
cells which were PMN was reduced to 53%. Forty seven percent of the cells
were mononuclear cells. The specific activity of PLA_2 in the extracellular
fluid also significantly increased from 79 nmol PL hydrolyzed/mg protein/hr a⁺

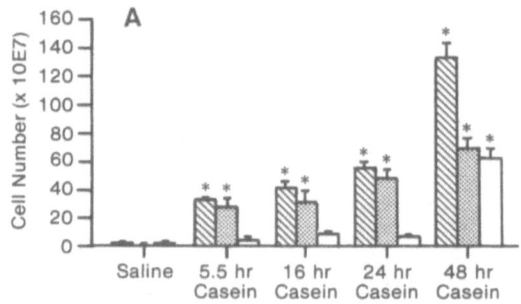

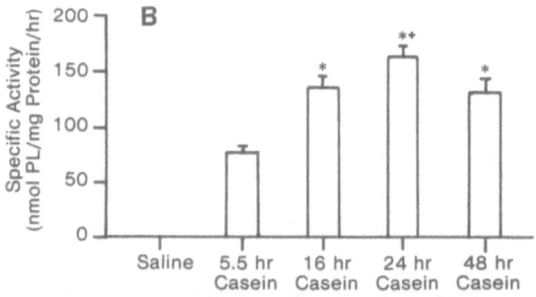

Fig. 5. (A). Time course of cell infiltration into the
peritoneal cavity of rats after casein injection. Total
cells and fluid were isolated from the peritoneal cavity at
5.5 hr, 16 hr, 24 hr or 48 hr after casein injection.
Total cells (◪), PMN (▥), and monocytes (□), were
determined. (B). Time course of appearance of PLA_2
activity in the lavage fluid was measured (□). * P < 0.05
compared to respective saline controls. + P < 0.05
compared to 5.5 hr, 16 hr and 48 hr. Mean ± SE, n = 4
rats/group.

5.5 hr after casein treatment up to 165 nmol PL hydrolyzed/mg protein/hr at 24 hr after casein injection. PLA_2 activity decreased 48 hr after casein treatment to 134 nmol PL hydrolyzed/mg protein/hr. Thus, although the total number of cells increased up to 48 hr, the peak of extracellular PLA_2 activity at 24 hr roughly paralleled the percentage of cells that were leukocytes in the peritoneal cavity. It should be noted that total cell influx may not have reached its peak by 48 hr.

DISCUSSION

The extracellular PLA_2 activities described in these two models of inflammation are similar: they are calcium dependent and have broad neutral pH optima. However, the timing of the cellular influx as it relates to PLA_2 activities are different. In zymosan peritonitis, the peak extracellular PLA_2 activity preceded the peak influx of cells by 6 to 8 hr; 8 hr after zymosan injection, the cellular influx plateaued. In contrast, the cells infiltrated the peritoneal cavity by 5.5 hr after casein injection and continued to increase through 48 hr. Extracellular PLA_2 activity was maximal within 15 min after injection of zymosan in the mouse model and peaked 24 hr after casein injection in the rat model, although times earlier than 5.5 hr after casein injection were not evaluated. Thus, the kinetics of cell influx and PLA_2 activities were different in these two models of inflammation. It should also be noted that casein treatment elicits 10 times more PLA_2 activity into the peritoneal cavity that does zymosan. The differences between these two models could be the result of species differences, or the use of two different inflammagens. However, PLA_2 activity in plasma is 10-fold that found in the peritoneal cavity in either model.

The tissue or cellular origin of PLA_2 activity in both the rat casein and mouse zymosan inflammation models has yet to be determined. We have shown previously and in this study that AA, one product of PLA_2 activation, appears within 30 min after injection of zymosan in mice and that plasma may be a primary source of both AA and PLA_2 activity[14]. Determination of the temporal relationship between PLA_2 activity and AA in rat casein inflammation awaits further experimentation. Several additional sources for PLA_2 activity may exist in both models. One possible source is release of PLA_2 by activated PMN in the peritoneal cavity. We have shown that the characteristics of the extracellular PLA_2 are similar to those of the cell-associated PLA_2 in the casein model. Another possible explanation suggested by Forst et al.[35] is that the PLA_2 in the blood is carried (i.e., "piggy-back") by the PMN migrating into the cavity. A third possible explanation is that the PLA_2 activity is derived from rat platelets in the plasma or present in the peritoneal cavity. Rat platelets are known to release extracellular PLA_2 in inflamed sites. The rat platelet-associated PLA_2 is structurally identical to the released extracellular rat platelet PLA_2 and is similar to the PMN-associated PLA_2[1,9]. Finally, activated resident mononuclear cells may serve as a source of PLA_2 activity.

In conclusion, extracellular PLA_2 activity has been found in "in vivo" models of inflammation, as well as rheumatoid arthritis, septic shock and pancreatitis. The source and contribution of this activity to the actual disease process(es) remain unclear. However, since this activity is present in so many diseases and models of disease, an antibody to extracellular PLA_2 or the development of a specific inhibitor may aid in understanding its importance.

REFERENCES

1. K. Horigome, M. Hayakawa, K. Inoue, and S. Nojima, Selective release of phospholipase A_2 and lysophosphatidylserine specific lysophospholipase from rat platelets, J. Biochem. 101:53 (1987).
2. M.D. Lister, K.B. Glaser, R.J. Ulevitch, and E.A. Dennis, Inhibition studies on the membrane-associated phospholipase A_2 in vitro and prostaglandin E_2 production in vivo of the macrophage-like P388D$_1$ cell. Effects of manoalide, 7,7-dimethyl-5,8-eicosandienoic acid and p-bromophenacyl bromide, J. Biol. Chem. 264:8520 (1989).
3. M.D. Lister, R.A. Deems, Y. Watanabe, R.J. Ulevitch, and E.A. Dennis, Kinetic analysis of the Ca^{+2}-dependent, membrane-bound, macrophage phospholipase A_2 and the effects of arachidonic acid, J. Biol. Chem. 263:7506 (1988).
4. P. Vadas, W. Pruzanski, J. Kim, and V. Fornasier, The proinflammatory effect of intraarticular injection of soluble human and venom phospholipase A_2, Am. J. Path. 134:807 (1989).
5. B.S. Vishwanath, A.A. Fawzy, and R.C. Franson, Edema-inducing activity of phospholipase A_2 purified from human synovial fluid and inhibition by aristolochic acid, Inflammation. 12:549 (1988).
6. W. Pruzanski, E.C. Keystone, B. Sternby, C. Bombardier, K.M. Snow, and P. Vadas, Serum phospholipase A_2 correlates with disease activity in rheumatoid arthritis, J. Rheumatol. 15:1351 (1988).
7. T.J. Nevalainen, The role of phospholipase A_2 in acute pancreatitis, Scand. J. Gastroenterol. 15:641 (1980).
8. P. Vadas and W. Pruzanski, Role of extracellular phopholipase A_2 in inflammation, Adv. Infl. Res. 7:51 (1983).
9. H.W. Chang, I. Kudo, S. Hara, K. Karasawa and K. Inoue, Extracellular phospholipase A_2 activity in peritoneal cavity of casein-treated rats, J. Biochem. 100:1099 (1986).
10. H.W. Chang, I. Kudo, M. Tomita, and K. Inoue, Purification and characterization of extracellular phospholipase A_2 from peritoneal cavity of caseinate-treated rat, J. Biochem. 102:147 (1987).
11. T.M. Stevens, M. McGowan, J. Giannaras and J.S. Kerr, Characterization of extracellular phospholipase A_2 activity in fluid and peritoneal cells from casein-treated rats, Inflammation (in press).
12. N.S. Doherty, P. Poubelle, P. Borgeat, T.H. Beaver, G.L. Westrich and N.L. Schrader, Intraperitoneal injection zymosan in mice induced pain, inflammation and the synthesis of peptidoleukotrienes and prostaglandin E_2, Prostaglandins 30:769 (1985).
13. K.R. Gans, S.R. Lundy, R.L. Dowling, T.M. Stevens and J.S. Kerr, Extracellular phospholipase A_2 activity in cell free peritoneal lavage fluid from mice with zymosan peritonitis, Agents Actions 27:341 (1989).
14. S.R. Lundy, R.L. Dowling, T.M. Stevens, J.S. Kerr, W.M. Mackin and K.R. Gans, Kinetics of phospholipase A_2, arachidonic acid and eicosanoid appearance in mouse zymosan peritonitis, J. Immunology (in press).
15. W. Pruzanski, P. Vadas, E. Stefanski, and M.B. Urowitz, Phospholipase A_2 activity in sera and synovial fluids in rheumatoid arthritis and osteoarthritis. Its possible role as a proinflammatory enzyme, J. Rheumatol. 12:211 (1985).
16. P. Vadas and J.B. Hay, The appearance and significance of phospholipase A_2 in lymph draining tuberculin reactions, Am. J. Pathol. 107:285 (1982).
17. R. Franson, C. Dobrow, J. Weiss, P. Elsbach, and W.B. Weglicki, Isolation and characterization of a phospholipase A_2 from an inflammatory exudate, J. Lipid Res. 19:18 (1978).

18. M. Waite. Phospholipase A_2 of mammalian cells. In: "Handbook of Lipid Research," Vol. 5, The Phospholipases. D.J. Hanahan, ed. Plenum Press, New York (1987).

19. S. Hara, I. Kudo, K. Matsuta, T. Miyamoto, and K. Inoue, Amino acid composition and NH_2-terminal amino acid sequence of human phospholipase A_2 purified from rheumatoid synovial fluid, J. Biochem. 104:326 (1988).

20. M. Hayakawa, I. Kudo, M. Tomita, and K. Inoue, Purification and characterization of membrane-bound phospholipases A_2 from rat platelets, J. Biochem 103:263 (1988).

21. S. Hara, I. Kudo, H.W. Chang, K. Matsuta, T. Miyamoto and K. Inoue, Purification and characterization of extracellular phospholipase A_2 from human synovial fluid in rheumatoid arthritis, J. Biochem. 105:395 (1989).

22. P. Patriarca, S. Beckerdite, and P. Elsbach, Phospholipases and phospholipid turnover in Escherichia coli spheroplasts, Biochem. Biophys. Acta. 260:593 (1972).

23. F.F. Davidson, E.A. Dennis, M. Powell, and J.R. Glenney, Inhibition of phospholipase A_2 by "lipocortins" and calpactin. An effect of binding to substrate phospholipids, J. Biol. Chem. 262:1698 (1987).

24. E.G. Bligh and W.J. Dyer, A rapid method of total lipid extraction and purification, Can. J. Biochem Physiol. 37:911 (1959).

25. G.R. Bartlett. Phosphorus assay in column chromatography, J. Biol. Chem. 234:466 (1959).

26. R. Franson, P. Patriarca, and P. Elsbach, Phospholipid metabolism by phagocytic cells. Phospholipase A_2 associated with rabbit polymorphonuclear leukocyte granules, J. Lipid Res. 15:380 (1974).

27. W.M. Mackin, S.M. Rakich, and C.L. Marshall, Inhibition of rat neutrophil functional responses by azapropazone, an anti-gout drug, Biochem. Pharmacol. 35:917 (1986).

28. A.A. Fawzy, R. Dobrow, and R.C. Franson, Modulation of phospholipase A_2 activity in human synovial fluid by cations, Inflammation 11:389 (1987).

29. B. Rothhut, F. Russo-Marie, J. Wood, M. DiRosa, and R.J. Flower, Further characterization of the glucocorticoid-induced antiphospholipase protein "renocortin," Biochem. Biophys. Res. Comm. 117:878 (1983).

30. M.M. Bradford. A rapid and sensitive method for quantitation of microgram quantities of protein utilizing the principle of protein-dye binding, Anal. Biochem. 72:248 (1976).

31. G.W. Snedecor. Sampling from normally distributed population. In: Statistical Methods, 5th ed., B.W. Snedecor and W.C. Cochran, eds. Iowa State Univ. Press Ames P. 45 (1964).

32. K. Burton, A study of the conditions and mechanisms of the diphenylamine reaction for the colorimetric estimation of deoxyribonucleic acid, Biochem J. 62:315 (1956).

33. K.W. Giles and A. Meyers, An improved dephenylamine method for the estimation of deoxyribonucleic acid, Nature 206:93 (1965).

34. S. Forst, J. Weiss, P. Elsbach, J.M. Maraganore, I. Reardon, and R.L. Heinrikson, Structural and functional properties of a phospholipase A_2 purified from an inflammatory exudate, Biochemistry 25:8381 (1986).

35. S. Forst, J. Weiss, and P. Elsbach, Properties of an inflammatory exudate phospholipase A_2 that degrades the phospholipids of E.coli during phagocytosis, Clin. Res. 34:724A (1986).

PHARMACOLOGICAL CONTROL OF PHOSPHOLIPASE A$_2$ ACTIVITY *IN VITRO* AND *IN VIVO*

Lisa A. Marshall and Joseph Y. Chang

Wyeth-Ayerst Research

Princeton, NJ

INTRODUCTION

Phospholipase(s)A$_2$ (PLA$_2$) are a class of enzymes that catalyze the hydrolysis of membrane phospholipids to liberate free fatty acids from the sn-2 position resulting in the formation of free fatty acids, predominantly arachidonic acid, and lysophospholipid. Within the last 20 years their role in disease has gained increasing attention (1,2,3). Various physiological stimuli (antigen-antibody complexes, cytokines, angiotensin II, bradykinin, prolactin and thrombin) activate PLA$_2$ when added to responsive cells. Indeed, high levels of PLA$_2$ activity have been found in joint fluid of patients with rheumatoid arthritis, in serum of patients with endotoxin shock or pancreatitis, in psoriatic lesions or peritoneal lavage fluids of patients with bacterial peritonitis. In some cases, enhanced levels of PLA$_2$ correlate with disease severity and have been proposed for use as a biochemical marker of disease activity (4,5).

Several laboratories, including our own, have established a clear relationship between the cytokine, interleukin-1 (IL-1) and PLA$_2$ induction in rabbit chondrocyte cultures, human synovial fibroblasts, and synovial fluid immune cells (6,7). Within the family of interleukins, IL-1 is unique in its ability to induce PLA$_2$ because other members of the family, such as IL-2 and IL-3, do not activate enzyme activity under similar conditions (8). Indeed in rabbit chondrocytes, tumor necrosis factor (TNF) was also unable to activate PLA$_2$. However, the mechanism of IL-1 induction of PLA$_2$ activity is unclear and the involvement of regulatory G proteins or protein kinase C is not well established. IL-1 may behave like a calcium ionophore and induces an influx of calcium which then increases PLA$_2$ activity or it may decrease the activity of or displace known endogenous PLA$_2$ inhibitors, such as lipocortin, from PLA$_2$. Nonetheless, direct demonstration that IL-1 induces a rapid rise in cell associated PLA$_2$ may, in part, explain some of the inflammatory and immune actions of IL-1.

Once activated, PLA_2 can mediate a variety of pathophysiological reactions either through a direct action or through subsequent transformation of its products (lysophospholipids and arachidonic acid) to several biologically active substances. For example, lysosphospholipids are cytotoxic substances and membrane fusagens and have been implicated in several human inflammatory conditions (9). Further, platelet activating factor (PAF), formed from 1-alkyl,2-lysophospholipid is a potent platelet aggregating substance and an inducer of various inflammatory reactions such as erythema, vascular permeability and cellular chemotaxis (10,11). Arachidonic acid-derived prostaglandins, thromboxane and leukotrienes formed by the cyclooxygenase (CO) or the 5-lipoxygenase (5-LO) enzyme systems are known for their multiple pro-inflammatory activities and complete the family of inflammatory and allergic mediators that can arise from PLA_2 activation (12). These facts provide a stronge rationale for a primary role of PLA_2 activation in the pathology of inflammation and for the notion that an inhibitor of this enzyme would provide a novel antiinflammatory therapy.

IDENTIFICATION AND DEVELOPMENT OF PLA_2 INHIBITORS

Numerous agents are reported to inhibit PLA_2 and represent a diversity of structural types. While such putative inhibitors are difficult to classify into distinct drug classes, it is nevertheless useful to place them within broad categories. As reviewed by Chang et al., these agents may be classified as agents that affect substrate-enzyme interface, agents that modulate calcium levels, non-steroidal antiinflammatory agents, natural products, covalent binding agents, substrate or product analogs and compounds derived from screening (3). Unfortunately, all these agents are limited in their *in vitro* potency and *in vivo* pharmacological activity. Apart from the *in vivo* studies with the steroid-induced PLA_2 inhibitory protein, lipocortin (13), most agents have not been useful in *in vivo* studies either due to the lack of poor specificity or oral bioavailability.

Judicious selection of *in vitro* and *in vivo* assays that demonstrate biochemical and antiinflammatory activity of proposed PLA_2 inhibitors is an important facet of PLA_2 research. Throughout this section, we will describe our approach that led to the identification of two chemically distinct PLA_2 inhibitors (WY-48,489 and WY-49,422, Fig. 1) that have provided some insight into the role of PLA_2 in the inflammatory process.

Enzymatic Identification of PLA_2 Inhibitors

At face value, determining the appropriate *in vitro* enzyme system appears straight forward; however, this selection should be given considerable thought . Two major objectives in choosing an appropriate assay system include: identification of a specific enzyme inhibitor and assay predictability of drug action *in vivo*. PLA_2 enzymes present special *in vitro* problems and their activity is significantly influenced by several co-factors and type of substrate and assay conditions.

A. Choice of Enzyme

It should be noted that the ubiquity and conservation of PLA_2 in all biological systems suggest that PLA_2 must serve a useful physiological function. There is evidence to indicate that membrane integrity, signal transduction, and food digestion are dependent, in part, to PLA_2 activity. Therefore, while the enhanced PLA_2 activity that is observed in inflammatory states provides a window of opportunity for pharmacological intervention, complete elimination of PLA_2 activity may not be desirable. With the isolation of various human enzymes, it is now possible to select the appropriate enzyme for drug screening. Total dependence on snake venom or pancreatic PLA_2 for pharmacological data is likely to produce PLA_2 inhibitors that do not exert an antiinflammatory effect. In the worst case scenario, such inhibitors may produce toxicity *in vivo* and, thereby, place a unwarranted stigma on future PLA_2 inhibitors that are developed through a more appropriate enzyme screen. Although molecular cloning of PLA_2 has been achieved and sequence homology identified, non-homologous structural components may contribute to enzyme function in which case functional isotypes may exist i.e. enzymes displaying distinct activity and/or functional differences. If functional PLA_2 isozymes exist, the opportunity to reduce potential side effects or liabilities through specific drug targeting may be possible.

In general, the molecular weights of purified PLA_2 from mammalian and non-mammalian sources are within the range of 12000 - 15000. The enzymes contain a high degree of disulfide cross-linking and are extremely heat stable and resistant to acid treatment; in fact, solubilization of membrane-bound enzymes can be easily achieved through mineral acid extraction. The isolation and purification of the PLA_2 protein has been reported for acylhydrolytic enzymes from human synovial fluid of patients with RA, human placenta, human platelet and neutrophil, and the serum of septic shock victums (14, 15).

The lack of a large source of relevant mammalian PLA_2 has been a major obstacle in fully understanding the role of PLA_2 in disease etiology. With the application of recombinant technology several laboratories are addressing this need. However until the proper expression system is delineated the isotype differences may be masked and go unrecognized. Currently, our laboratory is utilizing parallel evaluation of compounds *in vitro* by several PLA_2, representing a wide phylogenetic range. Perhaps, with time, these simultaneous screens that are run under identical conditions will generate the necessary data for answering the issues described above.

B. Choice of Substrate

Phospholipid substrate must be in an aggregated form for optimal presentation to the enzyme (16). PLA_2 will act on monomeric substrates following traditional Michaelis-Menten kinetics, but at an extremely low rate of hydrolysis. Upon reaching a substrate concentration where critical micellar mass is reached, phospholipids form aggregates and PLA_2 activity markedly increases. At this point, increases in substrate concentration are no longer representative of "available" substrate and classical kinetics no longer apply. Substrate sources are derived from two broad categories: (1) natural "membrane" lipid or (2) synthetic

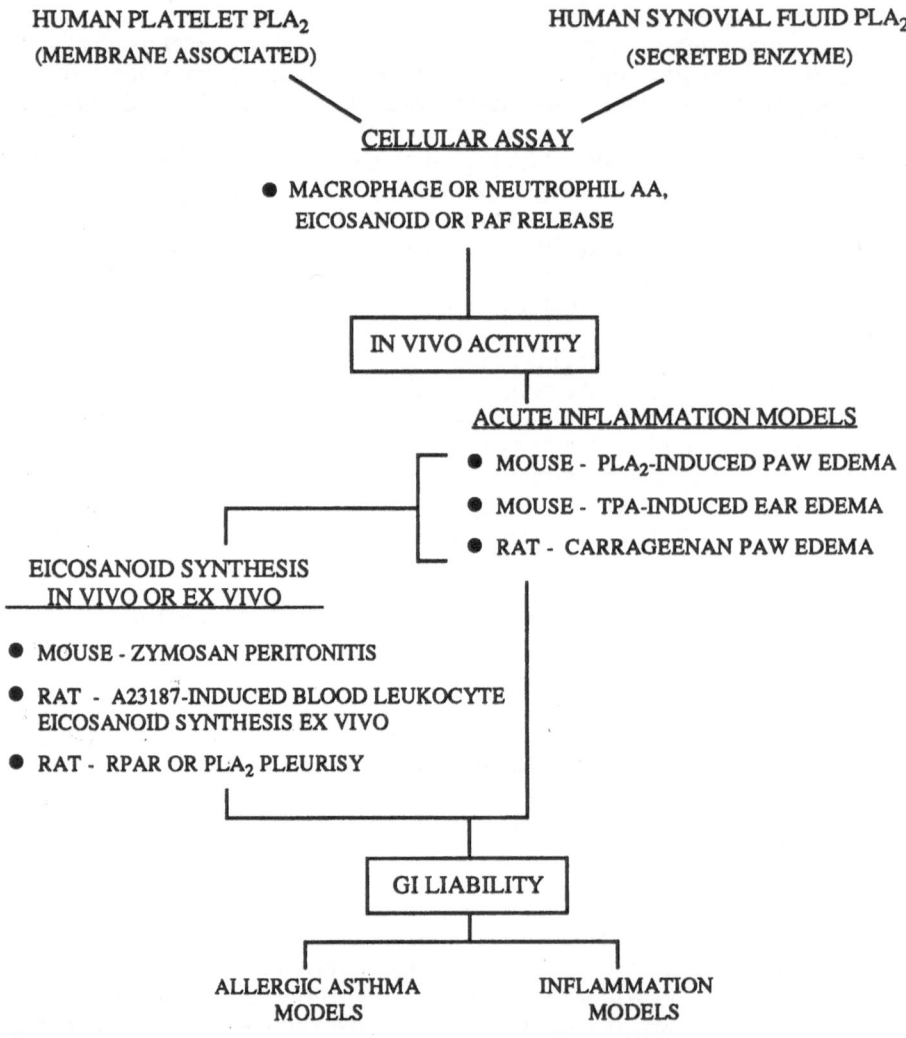

Fig. 1. DEVELOPMENTAL PATHWAY FOR ASSESSING
PHOSPHOLIPASE A₂ INHIBITORS

phospholipid forms (14,16,17). Membrane lipid composition is fairly complex and must be determined, whereas synthetic PL aggregates are of known composition. However, synthetic PL are by no means "natural lipid" presentation forms and display artificial lipid-aqueous interfaces and surface charges. Moreover, subtle effects of other membrane components (such as cholesterol or constitutive proteins) lipid domains, or bilayer asymmetry are not represented in the substrate milieu. The lipid aggregate forms used to assess PLA_2 activity include: monolayer dispersions, pure or combined lipid mixtures called liposomes, and detergent-lipid emulsions or micelles. Furthermore, the lipid aggregates can exist in various molecular forms, such as bilayer or hexagonal array. Parameters, including pH, ionic strength, calcium, temperature and type of PL, can all influence the physical form of the aggregate (14,16). Indeed, synthetic PL aggregates have a very low tolerance to changes in these assay conditions. The overall characteristics such as packing density and lipid surface charge can also have a marked effect on PLA_2 activation and hydrolysis.

Ultimately, the choice of the appropriate substrate depends on the target enzyme, level of sensitivity, throughput, correlation with *in vivo* data, and requirements for kinetic data.   The uncharacteristic enzyme behavior *in vitro* complicate the interpretation of the results and may interfere with extrapolation of data obtained by enzyme analysis to enzyme behavior in a biological system. Whether these factors will influence the interpretation of *in vitro* and *in vivo* pharmacological studies remains to be determined.

The Developmental Scheme for Identification and Development of PLA_2 Inhibitors

Taking the above issues into consideration, our laboratories has adopted as our initial primary screen an enzyme assay that employs both the membrane associated platelet PLA_2 (HP-PLA_2) and the extracellular "inflammatory" synovial fluid PLA_2 (HSF-PLA_2) (Fig. 2).   Essentially, these enzymes were chosen to bypass potential problems associated with species-related difference in enzyme activity. The selection of a membrane fatty acid-labeled phospholipid substrate (^{3}H-AA) labelled *E.coli*. (20) allows an approximation of the physiological location of phospholipids and may avoid problems associated with the inherent physicochemical characteristics of synthetic substrate as described above. With these assays, two chemical lead series were designed as transition state inhibitors e.g.glycidic esters (WY-49,422) and alkylamines (WY-48,489). Table 1 shows the inhibitory potencies of the two PLA_2 inhibitors compared to that of manoalide (MLD), a reported standard inhibitor derived from a natural marine source (18). WY-48,489 and WY-49,422 are less potent than MLD but display greater potency than other substrate analogues reported in the literature (14,18,19), such as the fluorinated phospholipid transition state analogues (IC_{50} = 0.07-1.6 mM). Table 1 also indicates that each compound inhibits various PLA_2 to varying degrees and highlights the difficulty of extrapolating *in vitro* data to *in vivo* systems. Even under identical conditions, IC_{50}'s varied with the type of enzyme and further investigation may reveal common or distinct catalytic functions of individual enzymes.

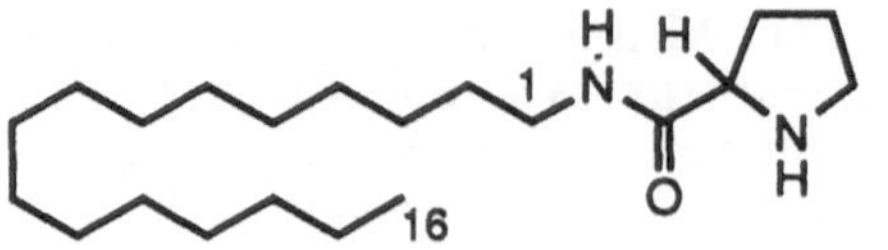

WY-48,489

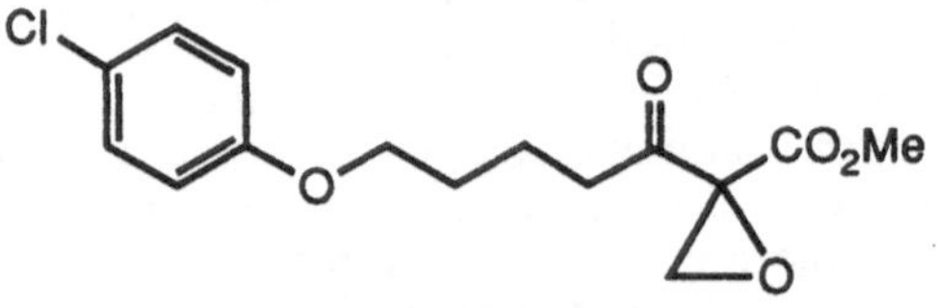

WY-49,422

Fig. 2. WYETH-AYERST PHOSPHOLIPASE A$_2$ INHIBITORS

MLD effectively inhibits all enzymes (IC$_{50}$ = 0.008 - 10.6 µM). The 1000-fold discrepancy in inhibitory activity between the different enzymes could be due in part to differences in the amount of non-enzyme related protein represented in each of the enzyme preparations. Indeed, reports exist demonstrating the "protective effect " of exogenous protein such as polylysine over PLA$_2$ when in the presence of manoalide (18,20) and may account for the lack of sensitivity in inhibiting the crude chondrocyte PLA$_2$ (IC$_{50}$ =3.3 µM). However, this is not the case with the action against highly purified *C. atrox* PLA$_2$ (IC$_{50}$ = 10.6 µM). This snake venom enzyme is known to exist as a dimer and may account in part for the lack of potent inhibitory action. While much work remains to be done to study this interesting finding, our data with manoalide are not entirely consistent with its proposed mechanism of action and suggest an additional, as yet undetermined, site(s) of action.

WY-49,422 was found to inhibit all enzymes at approximately the same relative potency (IC$_{50}$ = 2.3 - 13.0 µM). This action is unaffected by the variation in degree of purity between enzymes and infers that WY-49,422 interferes with a common component inherent in all PLA$_2$ enzymes. By contrast, WY-48,489 had no effect on the pancreatic, bee venom, or the snake venom PLA$_2$s; whereas it inhibited rabbit chondrocyte PLA$_2$, human platelet PLA$_2$, human synovial fluid PLA$_2$, in a descending order of potency. WY-48,489 therefore displayed selectivity for mammalian enzymes regardless of their state of purity suggesting the possible presence of inhibitor site-specific differences between mammalian and non-mammalian PLA$_2$. Taken together, these data demonstrate the possibility of designing selective PLA$_2$ inhibitors.

TABLE 1. COMPARATIVE POTENCIES OF PLA$_2$ INHIBITORS AGAINST VARIOUS PLA$_2$ ENZYMES

Drug	Human Platelet	Human Synovial Fluid	IL-1-induced Extracellular Chondrocyte	Pancreatic Lipase	Bee Venom	Snake Venom	
						C. atrox	A. p.p. (D-49)
			IC$_{50}$ (μM)				
MANOALIDE	0.31	0.02	3.3	0.024	0.008	10.6	0.007
WY-49,422	13.0	6.0	9.91	2.0	2.3	7.1	7.1
WY-48,489	62.7	119.5	29.9	NA	NA	NA	NA

Phospholipase A$_2$ (PLA$_2$) and drug were incubated for 1 h at 25°C in 0.2 M Tris buffer (pH 7.5) with 5 mM Ca^{++}. [^{3}H]AA E. coli (5 nmole phospholipid phosphorus) was added and mixed. The reaction mixture was incubated for 10 min at 37°C in a shaker bath. Two ml THF was added to stop the reaction. Free fatty acid was separated by aminopropyl solid phase column and [^{3}H]-AA was counted. All studies were performed at linear enzyme velocity conditions (% fatty acyl hydrolysis range from 4.8 - 16.7%). Enzymes were obtained from the following sources: Human platelet PLA$_2$ (123,000 fold pure; 0.07 μg protein/assay) and human synovial fluid PLA$_2$ (11,840 fold pure; 0.035 μg protein/assay) were column (3 fold) purified; extracellular rabbit chondrocyte PLA$_2$ from 24 h IL-1-treated cells (1.1 μg protein/assay), pancreatic PLA$_2$ (Sigma Co.; 1.6 ng protein/assay); bee venom PLA$_2$ (Sigma Co.; 2.4 ng protein/assay); snake venom C. atrox PLA$_2$ and D-49 A. p.p. were obtained from Dr. Paul Sigler, University of Chicago (12.5 ng and 0.48 ng protein/assay; respectively). NA = Not active at 100 μM

Assessment of PLA₂ Inhibitors: A Cellular Model

Our laboratory has also initiated several lines of investigation to determine the cellular and *in vivo* biological action of PLA$_2$ (Fig. 2). The majority of the data suggests that PLA$_2$ can initiate and propagate a series of events that are associated with the inflammatory reaction (21,22). For example, when macrophages were incubated in the presence of WY-48,489 and activated by A23187, AA release was indeed reduced with a concomitant decrease in both the major cyclooxygenase product, PGE$_2$ and 5-LO product, LTC$_4$ (Table 2). Interestingly, PLA$_2$ inhibition may not lead to a reduction in AA release under all conditions since the inhibitory effect varied depending on the stimulus. This observation is consistent with several studies where it has been demonstrated that the induction of AA release by TPA, zymosan, or calcium ionophore in macrophages is not identical (22). Similarly, MLD has also been shown to reduce AA release in mouse peritoneal macrophages only under certain conditions (23-25). From such data, we conclude that PLA$_2$ inhibitors that are active in a cell-free enzyme system will exert the expected action on AA mobilization by intact cells that are stimulated by the proper agonist.

Activity of PLA₂ Inhibitors in Classical Models of Inflammation

The most direct examination of *in vivo* activity of PLA$_2$ inhibitors is to utilize classical inflammation models which owe some aspect of their etiology to elevated eicosanoid levels. Such models include paw edema models, mouse zymosan peritonitis, reverse Arthus pleurisy or various skin inflammation assays (26). Table 3 shows the *in vivo* antiinflammatory activity of WY-48,489 and WY-49,422 in a paw and ear model of inflammation. Both compounds effectively reduced carrageenan paw edema when administered at either 50 or 100 mg/kg, p.o. Although the potency was not remarkable, WY-49,422 was at least equivalent in potency to ibuprofen. On the other hand, ear edema induced by topical application of the phorbol ester TPA (1 mg/ear) was effectively reduced by oral administration of WY-48,489 (ED$_{50}$ = 69 mg/kg) whereas the antiinflammatory effect of WY-49,422 was achieved by topical administration (-42% with 1 mg/ear). These data suggest that there are pharmacokinetic problems which require further chemical resolution before inhibitors of PLA$_2$ can be identified for clinical studies.

New Model Development

While classical models of inflammation will reflect the drug's ability to alter an inflammatory response, they cannot address the specificity of drug action. Clearly, these models have been traditionally designed as non-steroidal antiinflammatory drugs (NSAID)-sensitive pharmacological screens and it is important to identify additional models that can differentiate PLA$_2$ inhibitors from NSAIDs.

While it is self evident that PLA$_2$ activation leads to the synthesis of inflammatory lipid mediators, there is less appreciation that PLA$_2$ may cause direct inflammatory reactions related to the release of non-lipid mediators. In our hands, selective mediator antagonists or synthesis inhibitors are able to block only a portion of PLA$_2$-induced inflammation (27,28). Histamine and serotonin antagonists partially blocked PLA$_2$-induced edema even though these drugs did not

TABLE 2. EFFECT OF WY-48,489, A SELECTIVE PLA_2 INHIBITOR, ON MOUSE MACROPHAGE ARACHIDONIC ACID RELEASE, AND EICOSANOID GENERATION

Stimulating Agent	IC_{50} (μM)		
	^{3}H-Arachidonic Acid	PGE_2**	LTC_4**
A23187 (1 μM)	3.6 (3.0 - 4.4)	19.8 (13.5 - 28.9)	14.8 (11.6 - 18.9)
Zymosan (100 μg/ml)	19.6 (12.2 - 31.3)	13.17 (9.4 - 18.4)	7.56 (6.0 - 9.5)
TPA (100 nM)	11.9 (7.9 - 18.1)	11.2 (9.3 - 13.4)	No LTC_4 synthesized

IC_{50} values with the confidence limits were generated from nonlinear regression analysis of dose response curves.

* ^{3}H-AA measured from pre-radiolabelled mouse macrophage cultures after 2 h exposure to stimuli.

** PGE_2 and LTC_4 measured by radioimmunoassay in cultured mouse macrophage after 2 h exposure to stimuli. WY-48,489 preincubated with cells 5 min prior to addition of stimuli.

TABLE 3. PHARMACOLOGY OF WY-48,489 AND WY-49,422

Compound	IN VITRO *		IN VIVO	
	Human Platelet PLA$_2$ IC$_{50}$ (μM)	Human Synovial Fluid PLA$_2$	Rat Carrageenan Paw Edema Oral ED$_{50}$ (mg/kg)	TPA-induced Mouse Ear Edema
WY-48,489	16.1	119.5	>100	69
WY-49,422	18.4	36.3	~50	NA (Topical ED$_{50}$ ~1 mg/ear)
Ibuprofen	NA	NA	56	NA
Phenidone	NA	NA	NA	84

NA = Not active

IC$_{50}$s and ED$_{50}$s were calculated from dose-response curves that were analyzed by non-linear regression analysis.

* In vitro PLA$_2$ assays were performed under linear enzyme velocity conditions (10 - 12% hydrolysis of ^{3}H-AA labeled E. coli substrate).

inhibit PLA_2 in an enzyme assay (29). The PAF antagonists, CV 6202, and kadsurenone also partially inhibited the paw edema. Whereas CV 6202 did show some direct inhibitory activity against PLA_2 in an enzyme assay, kadsurenone was an ineffective PLA_2 inhibitor. High doses of steroids inhibited the edema and may be related to the induction of endogenous PLA_2 inhibitory proteins since there was a delay onset of action for this class of drugs. Finally, LO/CO inhibitors again were partially effective but were less active than the previous agents.

Inasmuch as the paw edema was induced by exogenously applied PLA_2, a PLA_2 inhibitor by interfering directly with the insult would be expected to block more effectively the inflammation. Indeed, prior inactivation of PLA_2 by p-bromophenacylbromide (pBPB) resulted in the marked inhibition of PLA_2 edema (> 80%), and luffariellolide and aristolocholic acid which inhibit human synovial fluid PLA_2 (29,30) also produced maximal inhibition when administered by co-injection. Two conclusions can be drawn from these studies. First, PLA_2 is proinflammatory and can be inhibited by PLA_2 inhibitors. The reaction is not an antigenic response since administration of other proteins, such as BSA (10 mg/paw) had no effect. Second, exogenously applied PLA_2 caused the release of a variety of mediators that contribute to the inflammatory reaction. The possible scenario derived from these pharmacological data is that after PLA_2 administration, degranulation of mast cells occurs followed by PAF and/or eicosanoid generation. No one selective mediator antagonist or inhibitor can totally abrogate the response, whereas a PLA_2 inhibitors is substantially more effective. PLA_2-induced paw edema therefore offers a well-defined whole animal model which can verify the *in vivo* efficacy of PLA_2 inhibitors.

Intrapleural injection of snake venom PLA_2 also induced a time- and dose-dependent inflammatory response in the pleural cavity (31). As early as one hour after PLA_2 administration fluid accumulation was significantly increased and continued to accumulate up to 24 hours. The appearance of neutrophils in the pleural exudate was observed by 2 hours and peaked at 6-24 hours. In addition, a reverse passive Arthus reaction (induced by the intravenous injection of BSA followed by 30 min later by the intrapleural injection of 1 mg antiBSA) caused endogenous PLA_2 release into the pleural cavity. In these studies, there was a gradual release of PLA_2 in the cavity increasing up through to four hours although this did not correlate with the early eicosanoid formation measured in the fluid (32,33). Further work is needed to understand the source and relationship of the eicosanoids, PLA_2 activity, cell migration and protein extravasation in this model. When fully defined, this model may provide the opportunity for examining the effects of PLA_2 inhibitors on endogenous PLA_2.

We have briefly described a set of studies that led to the design of a developmental scheme (Figure 2) that affords a rational evaluation of potential PLA_2 inhibitors. The proposed scheme appears to be satisfactory in terms of efficiency and generation of relevant *in vitro* and *in vivo* data. Although there are still many unanswered issues related to PLA_2 and the inflammatory process, the pharmacological models that we have described should provide further insight into the role of PLA_2 in inflammatory diseases. Already we have identified structurally novel compounds as PLA_2 inhibitors and with further modifications, more potent and orally active inhibitors should become available in the near future.

ACKNOWLEDGEMENT

We would like to thank Rosemary Clarke for her excellent secretarial assistance. Joseph Berkenkopf, William Calhoun, Dr. Richard Carlson, Lynn O'Neill-Davis and Dr. Barry Weichman are acknowledged for their collaboration on *in vivo* models and Eileen Blazek and Amy Sung for completing various *in vitro* analyses and Drs. Herb McGregor, John Musser, and Guy Schiehser for their chemical synthetic efforts.

REFERENCES

1. E. A. Dennis, Phospholipase A_2 Mechanism: Inhibition and Role in Arachidonic Acid Release, Drug Dev. Res. 10:205 (1987).
2. P. Vadas and W. Pruzanski, Biology of Disease: Role of Secretory Phospholipases A_2 in the Pathobiology of Disease, Lab. Invest., 55:391 (1986).
3. J. Chang, J. H. Musser and H. McGregor, Phospholipase A_2: Function and Pharmacological Regulation, Biochem. Pharmacol. 36:2429 (1987).
4. P. Vadas, Elevated Plasma Phospholipase A_2 Levels: Correlation with the Hemodynamic and Pulmonary Changes in Gram-Negative Septic Shock, J. Lab. Clin. Med. 104:873 (1984).
5. S. Forster, E. Ilderton, J. F. B. Norris, R. Summerly and H. J. Yardley, Characterization and Activity of Phospholipase A_2 in Normal Human Epidermis and in Lesion-Free Epidermis of Patients with Psoriasis or Eczema, Br. J. Dermatol. 112:135 (1985).
6. S. C. Gilman, J. Chang, P. R. Zeigler, J. Uhl and E. Mochan, Interleukin-1 Activates Phospholipase A_2 in Human Synovial Cells, Arthritis Rheum. 31:126 (1988).
7. W. Pruzanski, P. Vadas, J. Kim, H. Jacobs and E. Stefanski, Phospholipase A_2 Activity Associated with Synovial Fluid Cells, J. Rheumatol. 15:791 (1988).
8. J. Chang, S. C. Gilman and A. J. Lewis, Interleukin-1 Activates Phospholipase A_2 in Rabbit Chondrocytes: A possible Signal for IL-1 Action, J. Immunol. 136:1283 (1986).
9. J. N. Hawthorne and M. R. Pickard, Phospholipids in Synaptic Function, J. Neurochem. 32:5 (1979).
10. F. Snyder, M. L. Blank, D. Johnson, T. Lee, B. Malone, M. Robinson and D. S. Woodard, Alkylacetylglycerols Versus Lyso-PAF as Precursors in PAF Biosynthesis and the Role of Arachidonic Acid in PAF Metabolism, Pharmacol. Res. Commun., 18:33 (1986).
11. A. Etienne, F. Hecquet, C. Soulard, C. Touvay, F. Clostre and P. Braquet, The Relative Role of PAF-Acether and Icosanoids in Septic Shock, Pharmacol. Res. Commun., 18:71 (1986).
12. G. H. Higgs, B. Henderson, S. Moncada and J. A. Salmon, The Synthesis and Inhibition of Eicosanoids in Inflammation, in: "Inflammatory Mediators", G. A. Higgs and T. J. Williams, eds., VCH, Great Britain (1985).
13. R. J. Flower, Lipocortin and the Mechanism of Action of the Glucocorticoids, Br. J. Pharmacol. 94:987 (1988).
14. D. Mobilio and L. A. Marshall, Recent Advances in the Design and Evaluation of Inhibitors of Phospholipase A_2, in: "Annual Reports in Medicinal

Chemistry", Vol. 24, R. C. Allen, ed., Academic Press, Inc., New York (1989).

15. R. M. Kramer, C. Hession, B. Johansen, G. Hayes, P. McGray, E. P. Chow, R. Tizard, and R.B. Pepinsky, Structure and Properties of a Human Non-Pancreatic Phospholipase A_2, J. Biol. Chem. 264(10):5768 (1989).

16. A. Pluckthun and E.A. Dennis, Activation, Aggregation, and Product Inhibition of Cobra Venom Phospholipase A_2 and Comparison with Other Phospholipases, J. Biol. Chem. 260:11099 (1985).

17. P. Patriarca, S. Beckerdite, and P. Elsbach, Phospholipases and Phospholipid Turnover in *Escherichia coli* Spheroplasts, Biochim. Biophys. Acta 260:593 (1972).

18. K. B. Glaser and R. S. Jacobs, Inactivation of Bee Venom Phospholipase A_2 by Manoalide, Biochem. Pharmaco. 36:2079 (1987).

19. M. H. Gelb, Fluoro Ketone Phospholipid Analogues: New Inhibitors of Phospholipase A_2, J. Am. Chem. Soc. 108:3146 (1986).

20. C. F. Bennett, S. Mong, M. A. Clarke, L. I. Kruse, and S. T. Crooke, Differential Effects of Manoalide on Secreted and Intracellular Phospholipases, Biochem. Pharmacol. 36:733 (1987).

21. B. S. Vishwanath, A. A. Fauzy, R. C. Franson, Edema-inducing Activity of Phospholipase A_2 Purified From Human Synovial Fluid and Inhibition by Aristolochic Acid. Inflammation, 12:549 (1988).

22. J. L. Humes, S. Sadowski, M. Galavage, M. Goldenberg, E. Subers, R. J. Bonney, F.A. Kuehl, Jr., Evidence for Two Sources of Arachidonic Acid for Oxidative Metabolism by Mouse Peritoneal Macrophages, J. Biol. Chem. 257:1591 (1982).

23. A. M. S. Mayer, K. B. Glaser, and R. S. Jacobs, Regulation of Eicosanoid Biosynthesis *In Vitro* and *In Vivo* by the Marine Natural Product Manoalide: A Potent Inactivator of Venom Phospholipases, J. Pharmacol. Exp. Ther., 244:871 (1987).

24. M. Baggiolini, J. Schnyder, B. Dewald, U. Bretz, and T. G. Payne, Phagocytosis-Stimulated Macrophages, Production of Prostaglandins and SRS-A, and Prostaglandin Effects on Macrophage Activation, Immunobiol. 161:369 (1982).

25. L. A. Marshall, Arachidonic Acid Metabolism of Cultured Peritoneal Rat Macrophages and its Manipulation by Nonsteroidal Antiinflammatory Agents, Immunopharmacol. 15:177 (1988).

26 A. J. Lewis, R. P. Carlson, and J. Chang, Experimental Models of Inflammation, in: "Handbook of Inflammation" Vol. 5, I. L. Bonta, M. A. Bray, and M. J. Parnham, eds., Elsevier Science Publishers B.V., New York (1985).

27. L. A. Marshall, J.Y. Chang, W. Calhoun, J. Yu, and R. P. Carlson, Preliminary Studies on Phospholipase A_2-induced Mouse Paw Edema as a Model to Evaluate Antiinflammatory Agents, J. Cell. Biochem. 40:147 (1989).

28. G. Cirino, S. H. Peers, J. L. Wallace, and R. J. Flower, A Study of Phospholipase A_2-induced Oedema in Rat Paw, Eur. J. Pharmacol. 166:505 (1989).

29. M. R. Kerman and J. D. Faulkner, The Luffarrellins, Novel Antiinflammatory Sesteterpenes of Chemotaxononic Importance from the Marine Sponge Luffariella Vareabilis, J. Org. Chem. 52:3081 (1987).
30. W. Calhoun, J. Yu, B. M. Weichman, T. T. Chau, L. A. Marshall, and R. P. Carlson, Pharmacologic Modulation of Phospholipase A_2-induced Paw Edema in the Mouse, Agents Actions 27:418 (1989).
31. J. W. Berkenkopf, L. A. Marshall, B. M. Weichman, Phospholipase A_2-induced Inflammation in Rats, Tissue Reactions, in press.
32. J. W. Berkenkopf and B. M. Weichman, Differential Effects of Antiinflammatory Drugs on Fluid Accumulation and Cellular Infiltration in Reverse Passive Arthus Pleurisy and Carrageenan Pleurisy in Rats, Pharmacol. 34:309 (1987).
33. B. M. Weichman, J. W. Berkenkopf, C. A. Cullinan, and R. J. Sturm, Leukotriene B_4 Production and Pharmacologic Regulation of Reverse Passive Arthus Pleurisy: Importance of Antigen Dose, Agents Actions 21:351 (1987).

PHOSPHOLIPASE A$_2$ AS LEUKOTRIENE B$_4$ SECRETAGOGUE FOR

HUMAN POLYMORPHONUCLEAR LEUKOCYTES

Bing K. Lam[+], Chin-Yuh Yang[*] and Patrick Y-K Wong[**]

Departments of Physiology, Medicine
New York Medical College, Valhalla, N.Y.
and *Department of Dentistry, Tri-Service General Hospital
Taipei, Taiwan, R.O.C.

SUMMARY

High levels of soluble phospholipase A$_2$ (PLA$_2$) activity have been detected in tissues fluids associated with inflammatory diseases. However, the cellular origin for PLA$_2$ has not been demonstrated. Several groups of investigators have proposed that platelets, macrophages and chondrocytes may be the cellular source of this enzyme. In fact, soluble PLA$_2$ is secreted extracellularly from rabbit and rat chondrocytes and from human synovial cells in response to cytokine stimulation (1). PLA$_2$ activity has been shown to be increased upon stimulation by the chemotactic peptide (f-met-leu-phe) and thrombin in neutrophils and platlets (2). PLA$_2$ has been found to have pro-inflammatory effects and causes a dose dependent infiltration of leukocytes and increased vascular permeability (3). The vascular actions of PLA$_2$ have been proposed to be mediated through the release of prostaglandin E$_2$ and thromboxane (4). We have reported that purified PLA$_2$ from snake venom stimulated the release of leukotrienes and lipoxins from endogenous sources in porcine leukocytes. However, there is no information regarding the mechanism of action of human PLA$_2$ on inflammatory cells and the generation of leukotrienes. In this report, we present evidence that PLA$_2$ isolated from human platelets

[**]To whom all correspondence should be addressed
[+]Current Address:
Department of Rheumatol. and Immunol.
Brigham and Women's Hospital
Harvard Medical School
Boston, MA 02115

Phospholipase A2
Edited by P.Y.-K. Wong and E. A. Dennis
Plenum Press, New York, 1990

can stimulate the production of leukotriene B_4 from human polymorphonuclear leukocytes. These results suggest that soluble PLA_2 may function as a secretagogue of LTB_4 in inflammatory sites and further amplify the inflammatory processes by inducing chemotaxis of circulating leukocytes.

INTRODUCTION

Phospholipase A_2 (PLA_2) is one of the principal enzymes that liberates arachidonic acid (AA) from the SN-2 position of phospholipids (PLs). Once released, it can liberate arachidonic acid from phospholipids which are further metabolized to biologically active lipid mediators such as prostaglandins and leukotrienes. Phospholipase A_2 occurs in two forms: as a soluble, secretory enzyme and as a membrane associated non-secretory enzyme. Recent reports have demonstrated that secretory PLA_2 is associated with local and systemic changes in the production of autocoids. In addition, it may interact with cytokines in inflammatory diseases. High levels of PLA_2 activity have been found in synovial fluids of patients with inflammatory joint diseases. Recently, PLA_2 isolated from synovial fluids of patients with rheumatoid arthritis has been isolated, purified and characterized. The amino-acids sequence of one of the isoforms of this enzyme (peak A) had been determined and this particular form of PLA_2 had been cloned (5). To date, the mechanism of the release of this extracellular enzyme is still unknown. Recent evidences demonstrated that cytokines such as interleukin 1(IL-1) and tumor necrosis factor (TNF) can stimulate the release of this enzyme from chondrocytes and cultured rat calvaria osteoblasts (6). Thus the generation of cytokines during cell activations may be one of the primary events for the release of this enzyme that trigger the AA cascade and generation of inflammatory mediators.

The cardiovascular effects of the extracellular PLA_2 have been examined in different species and found to be pro-inflammatory (7). Most of its pro-inflammatory actions are related to the generation of prostaglandins E_2 (PGE_2) and thromboxane (TXB_2) (4). The extracellular PLA_2 from human platelets have been isolated and purified (8), but its physiological functions have not been investigated. In this report, we provide evidence that the extracellular PLA_2 isolated from human platelets can induce the release of leukotriene B_4 (LTB_4) from human leukocytes. Thus it is possible that PLA_2 released from platelets can function as "LTB_4 Secretagogue" during platelet and leukocyte interaction. The release of eicosanoids such as LTB_4 may further amplify the inflammatory action of PLA_2 at the site of injury by causing the infiltration of inflammatory cells.

MATERIALS AND METHODS

Preparation of human platelet phospholipase fraction:
Newly expired human platelet-rich plasma was obtained from a commercial source. Residual erythrocytes were separated by centrifugation at 120 x g for 10 minutes. The platelet-rich supernatant fluid was centrifuged at 1500 x g for 15 minutes to sediment the platelets. Platelets were resuspended in 5 ml of 50 mM Tris-HCl (pH 7.5) containing 1 mM ethylene glycol bis (β-aminoethylether)-N,N,N'-tetraacetic acid (EGTA) and sonicated for 1 minute. The sonicate was suspended centrifuged at 105,000 x g for 1 hr. at 4°C and then the supernatant was fractionated with ammonium sulfate to 50% saturation. After 1 hour at 4°C the solution was centrifuged at 10,000 x g for 10 min to precipitate the protein. The supernatant fluid was discarded and the pellet dissolved in 100 ml of 50 mM Hepes buffer.

Purification of PLA$_2$ From Human Platelets: The ammonium sulfate fraction was further purified by DEAE-cellulose and Sephadex G-100 column chromatography as described (9). The PLA$_2$ activity was monitored by the assay described above. By using this method we were able to purify the PLA$_2$ in human platelets by 5000 to 6000 folds (10). Preliminary results on the SDS-PAGE of this enzyme reviewed that the major band of protein had a molecular weight of 15000 KD (10). This partially purified enzyme was used in all experiments described in this report.

Phospholipase A2 assay: Each reaction mixture (100µl total volume) contained 50 mM Hepes (pH 7.0), 2mM CaCl$_2$ and 15µM 1-stearoyl-2-[^{14}C] arachidonoyl sn-Phosphotidylcholine (specific activity, 30mCi/mmol) which was dissolved in 10µl of dimethyl sulfoxide. The reaction was initiated by the addition of the crude enzyme or partial purified enzyme and incubated at 37°C for the 10 min (unless otherwise stated), and stopped by adding 500µl of ethanol containing 2% (vol/vol) acetic acid. Fifty µl of this mixture was applied to a silica gel TLC plate (channeled with preabsorbent zone, Analtech) and developed in ethyl acetate/acetic acid (99:1, vol/vol). One radioactive band was observed, migrating near the solvent front. This was removed from the plate and counted by liquid scintillation. In this solvent system the radiaoctive product co-migrates with authentic AA.

Cell preparation and Incubations: Human leukocytes were obtained from peripheral blood and prepared as previously described by Borgeat and Samuelsson (12). The cell preparations were contaminated with platelets (11) and mononuclear leukocytes. The leukocytes were suspended in Dulbecco's phosphate-buffered saline, pH 7.4, and diluted to 100 x 10^6 cells/ml. The viability of cells was confirmed by trypan blue exclusion and found to be greater than 90%. After preincubation of (10-20) x 10^8 leukocytes (10-20mL) for 5 min at 37°C in a shaking water bath, PLA2 was added and the incubation was continued for 10 min with continuous shaking. Viability of leukocytes after treatment with 10^{-9} to 3 x 10^{-7} M PLA_2 was determined by Trypan blue dye exclusion. Incubations were terminated by addition of ice-cold ethanol followed by immediate centrifugation at 300xg for 20 min. The ethanolic solution was evaporated under vacuum, and the residue was re-dissolved in 3mL of distilled water, acidified with 1N phosphoric acid to pH 3.5-5.0, and then extracted with nine volumes of ethylacetate. Ethylacetate extracts were dried under nitrogen: and subsequently dissolved in 50μl methanol. Separation was performed using RP-HPLC on a Waters Assocaited Dual Pump System equipped with an RP-HPLC ultrasphere ODS column (C_{18}-ODS, 5μ, 4.6mm x 25cm. Beckman, Palo Alto, CA), a U6-K injector and a 481 λ_{max} variable wavelength detector. The column was eluted with a linear gradient of methanol:water acetic acid (50:50:0.05 v/v) to methanol for 40 min at a flow rate of 1 mL/min. Column effluents were monitored at 270 nm (0-19min) and 235 nm (19-40 min), respectively. Following treatment with CH_2N_2, samples were repurified by an SP-HPLC as previously described (11).

U.V. Spectroscopy: Samples eluted from the HPLC were evaporated to dryness under vacuum, dissolved in absolute ethanol, and examined with a Hewlett-Packard 8450-A UV/VIA spectrophotometer.

Gas Chromatography-Mass Spectrometry: The methylesters of the materials recovered from the LTB_4 peak shown in Figure 1 were converted to trimethylisilyl ethers by addition of 25μL of pyridine followed by 50μL of hexamethyldisilazine (Supelco, Bellefonte, Pennsylvania). Mixtures were kept at room temperature for 20 min and dried under N_2. The sample was dissolved in 5μL hexane and injected into the gas chromatograph-mass spectrometer (Hewlett-Packard 5895-B) equipped with a glass column (0.5cm x 4m) packed with 1% SE-30 on Chromosorb W(HP), 80/100 mesh. Helium flow was set at 40mL/min, oven temperature was 200°C, injector temperature was 260°C, and ion source was 200°C. The electron energy was set at 70 eV (11).

RESULTS

    Incubations of PLA_2 partially purified from human platelets with PMN leukocytes induced the release of mono- and di- hydroxy eicosatetraenoic acids. The typical experiments yielded a total of five fractions after separation by RP-HPLC (Fig. 1). Fraction 3 coeluted with synthetic standard of LTB$_4$ (retention time of 27 min). After methylation, this fraction was further purified by SP-HPLC. UV spectrometer analysis of the methylester of this compound revealed a UV$_{max}$ of 270nm with shoulders at 260 and 280nm (Fig. 1 insert). GC/MS analysis of the TMSi derivatives of fraction 3 showed ions of high intensities at m/z: 203,217, and 383 while ions at low intensities were found at m/z: 293, 463,479 and 494, identical to the published mass spectrum of the methylester TMSi derivative of LTB$_4$ (12). Fraction 4 had a longer retention time (31.5 min) than LTB$_4$, U.V. spectrum of this function shows which UVmax 230nm, suggests it contains a conjugated double bond system. GC/MS analysis of the methylester TMSi derivative of the compound in fraction 4 gave a C-value of 23.5 and exhibited fragment ions at m/z: 496 (M), 481 (M-15), 465 (M-31), 406 (M-90), 203 (M-C$_1$-C$_5$), 213, 181, 131, 129, 119 and 105. This spectrum was similar to the published mass spectrum of 10,11-dihydro LTB$_4$ (13), suggested that fraction 4 is a metabolite of LTB$_4$ with one of its 3 conjugated double reduced. Similarly, by comparisons of the HPLC retention time, U.V. spectra and mass spectra to known standards, fractions 1, 2 and 5 were identified as 20-carboxyl LTB$_4$ (20-COOH LTB$_4$), 20-hydroxy-LTB$_4$ (20-OH LTB$_4$) and 5-HETE respectively (14).

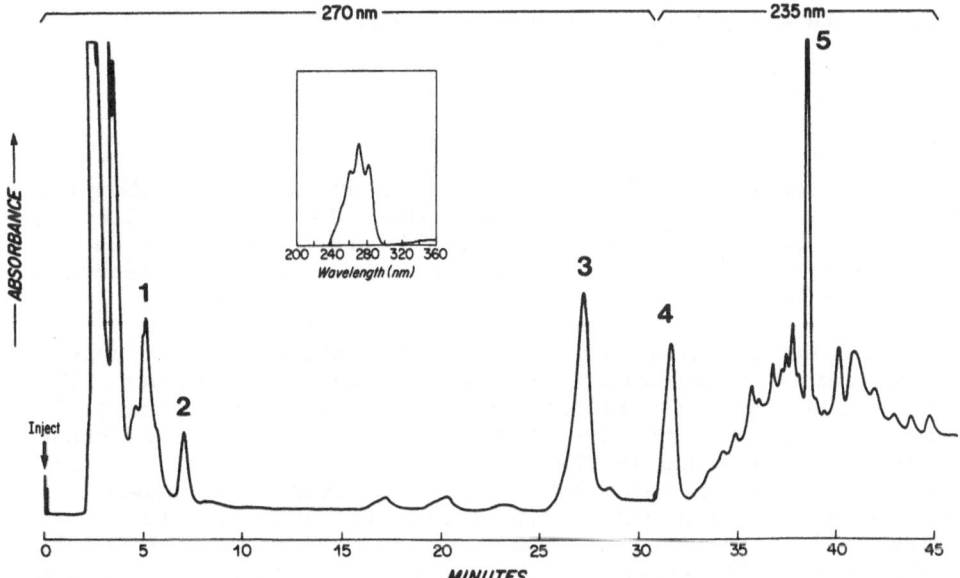

Figure 1. RP-HPLC chromatograms and U.V. spectrum of products extracted from incubation of human leukocytes after challenged with PLA$_2$ (1 x 10^{-7}M).

TABLE 1

CHROMATOGRAPHIC, ULTRAVIOLET AND GC/MS DATA OF LTB_4 AND ITS METABOLITES RELEASED FROM PMN LEUKOCYTES AFTER CHALLENGED WITH PLA_2

RP-HPLC PEAK	RETENTION TIME (mm)	UV_{max} (nm)	GC (C-VALUE)	MAJOR MASS IONS (m/z)	DESIGNATIONS
1	5	270	26.5	203,383 523,248	20-COOH LTB_4
2	7	270	25.4	203,301 507,582	20-OH LTB_4
3	27	270	23.8	203,213 479,494	LTB_4
4	31.5	232	31.5	203,481,496	10,11-DIHYDRO LTB_4
5	38.2	234	21.4	203,255,406	5-HETE

Reverse phase (RP) HPLC peak designations correspond to Figure 1. HPLC retention times can occur as column activity changes. GC (CV) refers to the elution time of the $Me-Me_3$ Si (TMSi) derivations relative to a series of saturated fatty acid Me standards. MS ions are representative fragment ions in the mass spectrum. UV_{max} denotes the wavelength of maximum absorption.

DISCUSSION

In this report we demonstrated that human PMN leukocytes released LTB_4 and three major metabolites after being challenged with platelet PLA_2. The generation of LTB_4 and its metabolites may be due to the mobilization of arachidonic acid from membrane phospholipids by PLA_2.

We had reported that snake venom PLA_2 isoenzyme is a potent stimulant for the generation of lipoxins and leukotrienes from endogenous sources of arachidonic acid in porcine leukocytes (15). However, the exact mechanism of how PLA_2 acts on the membrane phospholipids that leads to the generation of eicosanoids is still unknown. The large amounts of LTB_4 released from human PMN leukocytes suggest that intracellular PLA_2 released from platelets may play an important role in inflammation. PLA_2 is released by macrophages and leukocytes during cell activation and during platelet aggregation. High PLA_2 activities are present at the site of inflammatory joint disease (3). Most importantly, PLA_2 has been shown to mediate inflammatory hyperemia (3,4). It is likely that

Inflammatory Cells

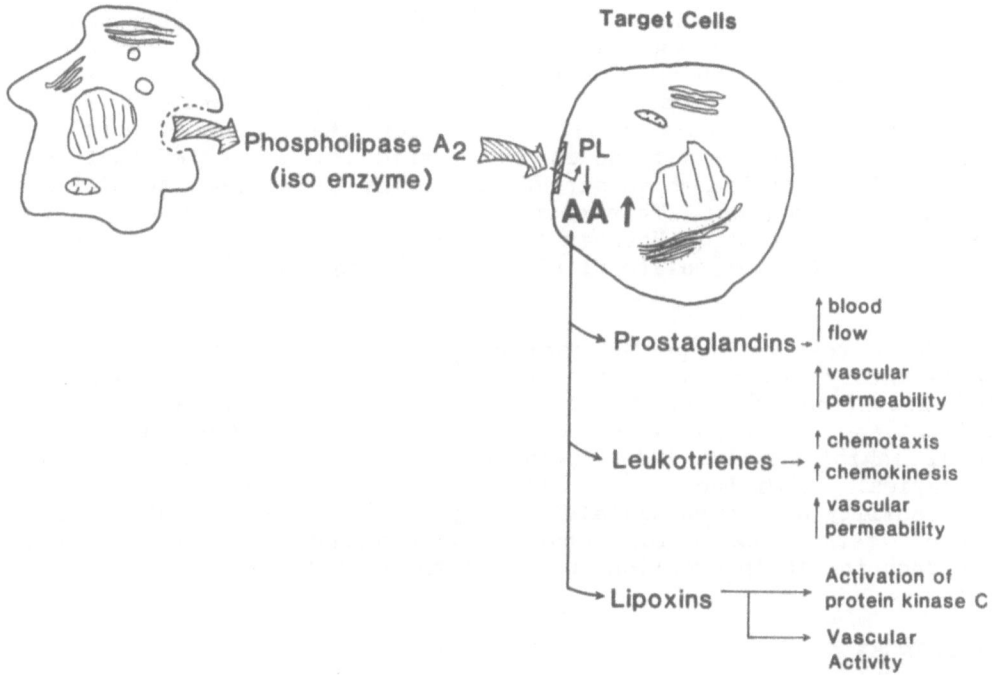

Figure 2.   Proposed mechanism for PLA$_2$ stimulated release of arachidonic acid derived lipid mediators.

activated platelets release PLA$_2$ at the site of inflammation or during platelet lekuocytes interaction (15). Arachidonic acid liberated from membrane phospholipids could be metabolized by cyclooxygenase to form prostaglandins or via the 5-lipoxygeanse to yield leukoctrienes (Fig. 2).   Prostaglandins will regulate the local vascular activities and permeability, while LTB$_4$, the most potent chemotactic substance so far identified, will recruit more leukocytes to the site of inflammation and further amplify the inflammatory process.   Under our experimental conditions, human mixed leukocytes produced a complete profile of 5-lipoxygenase products including LTB$_4$ and its metabolites i.e., 20-OH LTB$_4$, 20-COOH LTB$_4$ and 10,11-dihydro LTB$_4$ respectively.   The detection of large amount of 5-HETE in the incubations of PMN leukocytes with PLA$_2$ suggested that the 5-lipoxygenase may also be activated.   Our results demonstrated that after interaction with PLA$_2$, human leukocytes not only generated LTB_4 but also contained enzymes that metabolized LTB_4 to three different compounds: 20-COOH LTB_4, 20-OH LTB_4 and 10,11-dihydro LTB_4. Control experiments of incubations with mixed leukocytes alone failed to release any detectable amount of LTB_4 and its metabolites by RP-HPLC. Powell and coworkers reported that in human granulocytes, authentic LTB_4 was not a substrate for reduction, but was exclusively transformed to ω and ω-1 oxidative products (13). The fact that large amount of 10,11-dihydro LTB_4 was found in our experiments, suggested that the LTB_4 10,11-reductase is an active enzyme for the metabolism of LTB_4 in human leukocytes. This enzyme may be modulated by exogenous factors such as PLA_2. However the biological activity and the role of this dihydro metabolite of LTB_4 in inflammation still remains to be defined.

These results suggested that PLA_2 exerts an important role in inflammatory diseases. Therefore the regulation of the synthesis and secretion of PLA_2 from inflammatory cells may control the initiation and propagation of inflammation. The development of novel inhibitors for PLA_2 would provide new approaches for drug therapies. With the recent development in molecular biology and availability of cloned platelet and synovial fluid PLA_2 make PLA_2 a perfect target enzyme for future drug development and therapeutic approach for the prevention and treatment of inflammatory diseases.

REFERENCES

1. Chang, J., Gilman, S.C. and Lewis, A.J. Interleukin-1 activates phospholipase A_2 in rabbit chondrocytes: A possible signal for IL-1 action, J. Immunol. 136:1283, (1986).
2. Vadas, P., Pruzanski, W., Stefanski, E., Ellies, L., Aubin, J. Interleukin-1 induced syntehsis and secretion of extracellular phospholipase A_2 from the cultured bone forming cells, International Conference on "Lipoproteins and Phospholipases", Paris, France, September, (1988).
3. Purzanski, W. Experimental evidence for a proinflammatory effect of phospholipase A_2 on joint tissue. J. Rheumat. 13, (1990), 1986.
4. Huang, H.C. Effects of phospholipases A_2 from vipera russelli snake venom on blood pressure, plasma prostacyclin level and renin activity in rats, Toxicon, 22:253, (1984).
5. Seilhamer, J.J., Pruzanski, W., Vadas, P., Plant, S., Miller, J.A., Kloss, J., Johnson, L.K. Cloning and recombinant expression of phospholipase A_2 present in rheumatoid arthritic synovial fluid, J. Biol. Chem. 264: 5355, (1989).
6. Vadas, P., Pruzanski, W., Stefanski, E., Ellies, L., Aubin, J., IL-1 induces the synthesis and secretion of a soluble phospholipase A_2 from fetal rat calvarial bone cells, Sclavo International Conference, (1989).

7. Seilhamer, J., Vadas, P., Purzanski, W., Plant, S., Stefanski, E. and Johnson, L.   Synovial fluid phospholipase A_2 in arthritis, in Therap. Approaches to inflammatory diseases. Ed. by J. Lewis, N.S. Doherty and N.R. Ackerman. Elsevier Science Press, Inc., NY, pp. 129, 1989.

8. Kramer, R.M., Hession, C., Johansen, B., Hayes, G., McGray, P., E.P. Tizard, R., Pepinsky, R.B. Structure and properties of a human non-pancreatic phospholipase A_2.   J. Biol. Chem. 264, 5768, (1989).

9. Wong, P.Y-K, Lee, W., Chao, P.H-W, Reiss, R.F. and McGiff, J.C. Metabolism of prostacyclin by 9-hydroxyprostaglandin dehydrogenase in human platelets. J. Biol. Chem. 255:9021, (1980).

10. Parks, T.P. and Wong, P.Y-K.   Purification of expoxygenase/ phospholipase A_2 in human platelets, manuscript in preparation, (1990).

11. Wong, P.Y-K, Westlund, P., Hamberg, M., Granstrom, E., Chao, P.H-W and Samuelsson, B. ω-Hydroxylation of 12-HETE in human polymorphonuclear leukocytes.   J. Biol. Chem. 259: 2683, (1984).

12. Borgeat, P. and Samuelsson, B.   Arachidonic acid metabolism in polymorphonuclear leukocytes; effects of ionophore A23187, Proc. Natl. Acad. Sci. USA 76:2148, (1979).

13. Powell, W.S. and Gravelle, F.   Metabolism of leukotriene B_4 to dihydro- and dihydro-oxo-products by procine leukocytes. J. Biol. Chem. 264: 5364, (1989).

14. Samuelsson, B., Leukotrienes.   Mediators of immediate hypersensitivity reactions and inflammation, Sciences 220:568, 1983.

15. Lam, B.K., Serhan, C.N., Samuelsson, B. and Wong, P.Y-K.   A phospholipase A_2 isoenzyme provokes lipoxin B formation from endogenous sources of arachidonic acid in porcine. Leukocytes. Biochem. Biophys. Res. Commun., 144: 123, 1987.

CONTRIBUTORS

C. Anderson
Department of Allergy & Inflammation
Hoffmann-LaRoche, Inc.
Nutley, NJ   07110

John S. Bomalaski
Veterans Administration Medical Ctr.
Medical College of Pennsylvania
University of Pennsylvania

Joseph Y. Chang
Wyeth-Ayerst Research
Princeton, NJ

Mike A. Clark
Schering-Plough Research
Bloomfield, NJ

J. Coffey
Department of Allergy & Inflammation
Hoffmann-LaRoche, Inc.
Nutley, NJ   07110

Theresa M. Conway
Smith Kline and French
Philadelphia, PA

Mike Cook
Smith Kline and French
Philadelphia, PA

Stanley T. Crooke
ISIS
San Diego, CA

Edward A. Dennis
Department of Chemistry
University of California at San Diego
La Jolla, California   92093

Janice Dispoto
Veterans Administration Medical Center
Medical College of Pennsylvania
University of Pennsylvania

Randine L. Dowling
E.I. duPont de Nemours & Co.
Medical Products Department
P.O. Box 80400
Wilmington, DE 19880-0400

Susan Frank
The Depts. of Medicine & Microbiology
Jonsson Comprehensive Cancer Center
UCLA
Los Angeles, CA   90024

Kathleen R. Gans
E.I. duPont de Nemours & Co.
Medical Products Department
P.O. Box 80400
Wilmington, DE 19880-0400

Mary E. Gerritsen
Department of Physiology
New York Medical College
Valhalla, NY 10595

Keith B. Glaser
Department of Chemistry
University of California at San Diego
La Jolla, CA 92093

Catherine Hession
Biogen Inc.
Cambridge, MA

Ann F. Hoffman
Department of Biochemistry
Boehringer Ingelheim Pharmaceuticals
90 East Ridge
Ridgefield, CT   06877

Lorin K. Johnson
Salix Pharmaceuticals
1507 Kennewick Drive
Sunnyvale, CA 94087

Berit Johansen
Norwegian Institute of Technology
Trondheim, Norway

Janet S. Kerr
E.I. duPont de Nemours & Co.
Medical Products Department
P.O. Box 80400

Ruth M. Kramer
Lilly Research Laboratories
Indianapolis, IN

Bing K. Lam
Department of Rheumatology & Immunology
Brigham and Women's Hospital
Boston, MA   02115

Mark D. Lister
Department of Chemistry
University of California at San Diego
La Jolla, California   92093

Susan Lukas
Department of Biochemistry
Boehringer Ingelheim Pharmaceuticals
90 East Ridge
Ridgefield, CT   06877

Susan R. Lundy
E.I. duPont de Nemours & Co.
Medical Products Department
P.O. Box 80400
Wilmington, DE 19880-0400

Aldons J. Lusis
The Depts. of Medicine & Microbiology
Jonsson Comprehensive Cancer Center
UCLA
Los Angeles, CA   90024

William M. Mackin
E.I. duPont de Nemours & Co.
Medical Products Department
P.O. Box 80400
Wilmington, DE 19880-0400

Robert J. Mannix
Department of Physiology
New York Medical College
Valhalla, NY 10595

Lisa A. Marshall
Wyeth-Ayerst Research
Princeton, NJ

K. Meyers
Department of Allergy & Inflammation
Hoffmann-LaRoche, Inc.
Nutley, NJ   07110

Seymore Mong
Smith Kline   and French
Philadelphia, PA

K. Moody
Department of Allergy & Inflammation
Hoffmann-LaRoche, Inc.
Nutley, NJ 07110

D.W. Morgan
Department of Allergy & Inflammation
Hoffmann-LaRoche, Inc.
Nutley, NJ 07110

Thomas P. Parks
Department of Biochemistry
Boehringer Ingelheim Pharmaceuticals
90 East Ridge
Ridgefield, CT 06877

R. Blake Pepinsky
Biogen, Inc.
Cambridge, MA

Waldemar Pruzanski
Immunology Diagnostic and
  Research Center
Department of Medicine
Wellesley Hospital
University of Toronto
Toronto, Ontario
Canada M4Y 1J3

Jan Marian-Scardina
California Biotechnology Inc.
2450 Bayshore Parkway
Mountain View, CA 94043

Jeffrey J. Seilhamer
Ideon Corporation
515 Galveston Drive
Redwood City, CA 94063

Robert G.L. Shorn
AT Biochem
Malvern, PA

Jeff Stadell
Smith Kline and French
Philadelphia, PA

Theresa M. Stevens
E.I. duPont de Nemours & Co.
Medical Products Department
P.O. Box 80400
Wilmington, DE 19880-0400

Richard J. Ulevitch
Department of Immunology
University of California at San Diego
La Jolla, California 92093

Peter Vadas
Immunology Diagnostic and
  Research Center
Department of Medicine
Wellesley Hospital
University of Toronto
Toronto, Ontario
Canada M4Y 1J3

Lynne Webb
Schering-Plough Research
Bloomfield, NJ

A. Welton
Department of Allergy & Inflammation
Hoffmann-LaRoche, Inc.
Nutley, NJ 07110

Patrick Y-K Wong
Department of Physiology and Medicine
New York Medical College
Valhalla, NY 10595

Chin-Yuh Yang
Department of Dentistry
Tri-Service General Hospital
Taipei, Taiwan, R.O.C.

Grace Wright
New York University School of Medicine
Departments of Microbiology & Medicine
New York, New York

INDEX

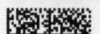